End-to-End Data Analytics for Product Development

End-to-End Data Analytics for Product Development

A Practical Guide for Fast Consumer Goods Companies, Chemical Industry and Processing Tools Manufacturers

Rosa Arboretti
University of Padova,
Italy

Mattia De Dominicis
Reckitt Benckiser,
Italy

Chris Jones
Reckitt Benckiser,
USA

Luigi Salmaso
University of Padova,
Italy

This edition first published 2020
© 2020 John Wiley & Sons Ltd

Registered Offices
John Wiley & Sons, Inc., 111 River Street, Hoboken, NJ 07030, USA
John Wiley & Sons Ltd, The Atrium, Southern Gate, Chichester, West Sussex, PO19 8SQ, UK

Editorial Office
9600 Garsington Road, Oxford, OX4 2DQ, UK

For details of our global editorial offices, customer services, and more information about Wiley products, visit us at www.wiley.com.

Wiley also publishes its books in a variety of electronic formats and by print-on-demand. Some content that appears in standard print versions of this book may not be available in other formats.

Library of Congress Cataloging-in-Publication data applied for
HB ISBN: 9781119483694

Cover Design: Wiley
Cover Image: © polygraphus/iStock.com

Set in 10/12pt WarnockPro by SPi Global, Chennai, India
Printed and bound by CPI Group (UK) Ltd, Croydon, CR0 4YY

10 9 8 7 6 5 4 3 2 1

Contents

Biographies

Rosa Arboretti, PhD, is Associate Professor of Statistics at the Department of Civil, Environmental and Architectural Engineering of the University of Padova. She has had over 100 papers and books published. Her main research interests are in Design of Experiments, Biostatistics, Pharmaceutical Statistics, Nonparametric Statistics, Customer Satisfaction Surveys, Machine Learning and Big Data Analytics.

Mattia De Dominicis is a former R&D Vice-President in Reckitt Benckiser with a successful track record in R&D, having led innovation, product development, launch and roll out of consumer good products in Household and Personal Care. He has lived and worked in Italy, the US, UK, South Africa, France and Germany and has experience in inspiring multicultural teams to outperform through fresh vision. Mattia is experienced in growing brands through product innovation and in managing complex projects. He graduated from the University of Padova.

Chris Jones is a Vice President of R&D in Reckitt Benckiser leading product development for key power brands. Chris has lived and worked in the US, Italy and the UK and developed products across Household and Personal Care sectors whilst striving to develop winning teams and advancement of his personnel. Chris holds a PhD in organometallic chemistry from the University of Liverpool.

Luigi Salmaso, PhD, is Full Professor of Statistics at the Department of Management and Engineering of the University of Padova and Deputy Chair of same Department. He has had over 300 papers and books published. His main research interests are in Design of Experiments, Six Sigma, Quality Control, Nonparametric Statistics, Industrial Statistics, Machine Learning and Big Data Analytics.

Preface by **Chris E. Housmekerides**, Senior Vice President R&D Reckitt Benckiser, Head of Innovations & Operations, Hygiene Home. Chris holds a PhD in inorganic chemistry from Penn State University. He has lived and worked in Belgium, China, Germany, Italy, the Netherlands and the US, published a plethora of patent applications in FMCGs, with R&D workstream leadership in M&As, R&D reorganization/restructuring/change management. He is passionate about R&D in the hygiene and homecare sector. He spent almost three decades questioning how industries can be more innovative with consumer products, challenging brands to provide better solutions, for today and tomorrow. His dream for the future is to influence through grass roots education – helping leaders of the future understand how to save energy, water, and lives through purpose-led products.

Preface

The fast-moving consumer goods (FMCG) industry includes a variety of sectors, from household products to food and beverages to cosmetics and over-the-counter (OTC). Despite products being very different in terms of their intended use, benefits, and the way they are regulated, there is a common approach to the method of developing them and identifying the solution for consumers. The process of innovation is similar: starting from the consumer understanding, identifying the insights that are most relevant, and then proceeding to develop an idea. A solution can solve a problem initially at the conceptual level, then moving across to an actual prototype, and then finally to new product development (NPD). No matter the segment, the appropriate use of data presents a value-adding opportunity to NPD.

There is a great focus in the industry around using big data derived from multiple sources, which are easily accessible via the internet. Today, information derived from many global applications is increasingly measuring different aspects of our lives. However, the focus is now to gain business value from such data. Surprisingly, the use of data for product development in most cases is not robustly established; a clear connection between the experts in statistics/data analytics and the product developers within the main FMCG companies needs to be established.

The focus is to develop better products for consumers by investing the right features and to support faster developments to ultimately limit the number of experiments. So, the scope of this book is to demonstrate how actual statistics, data analytics and design of experiment (DoE) can be widely used to gain vital advantages with respect to speed of execution and lean formulation capacity. The intention is to start from feasibility screening, formulation, and packaging development, sensory tests, etc. introducing relevant techniques for data analytics and guidelines for data interpretation. Process development and product validation can also be optimized through data understanding, analysis and validation.

Throughout the book, there is an intentional balance between the discussion of real case studies and topic explanations made as clear and intuitive as possible. Details on statistical and analytical tools are presented in sections denoted as Stat Tools. Each case study includes a guide to apply the proposed techniques with Minitab, a widely used general-purpose statistical software (Cintas, P. G., et al., 2012). Some examples with JMP software are also developed as online supplementary material.

This book is an exciting collaboration between expert Statisticians from the University of Padova and innovators/product developers from Reckitt Benckiser (RB) over more than 15 years. The passion and the involvement of multiple people within RB and the University of Padova in various projects have allowed for the development of the examples discussed in this book. A big thanks goes to the leading management group and all people from RB who contributed directly or indirectly in giving inspiration to write this book.

Dr. Chris Housmekerides
SVP R&D RB
Head of Innovations & Operations, Hygiene Home

About the Companion Website

This book is accompanied by a companion website:

www.wiley.com/go/salmaso/data-analytics-for-pd

The website includes:

– Case studies
– Different projects with JMP software

Scan this QR code to visit the companion website.

1

Basic Statistical Background

1.1 Introduction

Statistics and data analytics play a central role in improving processes and systems and in decision-making for strategic planning and manufacturing (Roberts, H. V., 1987). During experimental research, statistical tools can allow the experimenter to better organize observations, to specify working hypotheses and possible alternative hypotheses, to collect data efficiently, and to analyze the results and come to some conclusions about the hypotheses made.

This is an introductory chapter where readers can review several basic statistical concepts before moving on to the next chapters. Sixteen sections titled Stat Tools will introduce some key terms and procedures that will be further elaborated and referred to throughout the text.

Specifically, this chapter deals with the following:

Topics	Stat tools
Statistical variables and types of data	1.1
Statistical Units, populations, samples	1.2
Introduction to descriptive and inferential analyses	1.3, 1.12, 1.13
Data distributions	1.4, 1.5
Mean values	1.6, 1.7
Measures of variability	1.8, 1.9, 1.10
Boxplots	1.11
Introduction to confidence intervals	1.14
Introduction to hypothesis testing procedures, including the p-value approach	1.15, 1.16

Learning Objectives and Outcomes

Upon completion of the review of these basic statistical concepts, you should be able to do the following:

Recognize and distinguish between different types of variables.
Distinguish between a population and a sample and know the meaning of random sampling.
Detect the shape of data distributions.
Calculate and interpret descriptive measures (means, measures of variability).

Understand the basic concept and interpretation of a confidence interval.
Understand the general idea of hypothesis testing.

Understand the p-value approach to hypothesis testing.

Stat Tool 1.1   Statistical Variables and Types of Data

In statistical studies, several characteristics are observed or measured to obtain information on a phenomenon of interest. The observed or measured characteristics are called *statistical variables*. Statistical variables differ according to the type of values they store.

Qualitative or *categorical* variables can assume values that are qualitative categories and can be either ordinal or nominal.

Quantitative or *numeric* variables can assume numeric values and can be discrete or continuous. Discrete data (or count data) are numerical values only measurable as integers. Continuous data are numeric values (typically instrumental measures) that can be meaningfully subdivided into fractions.

Type of variables

Qualitative or *categorical*
assume qualitative values
(e.g. high, low, good, bad, ...)

Quantitative or *numeric*
assume quantitative values
(e.g. 12%, 3°, 350 m, ...)

Ordinal/ordered
ordered categories
(e.g. low, high)

Nominal/unordered
categories NOT ordered
(e.g. green, white)

Discrete
whole numbers
(e.g. 0, 1, ... defects)

Continuous
may contain
decimals
(e.g. 45.8g)

➤ *Example 1.1*. For a new shaving oil, it is of interest to compare fragrance A and fragrance B to investigate which is preferred. A sample of female respondents is presented with the two fragrances and asked their age and to answer the following question: How suitable or unsuitable is the fragrance to a shaving aid? Each was asked to assign an integer score from 0 (very unsuitable) to 10

Stat Tool 1.1 (Continued)

(very suitable). Respondents also expressed their purchase intentions for selecting one of the following categories: Probably would not buy it, Neither, Probably would buy it.

The variable "Fragrance" is a nominal categorical variable, assuming two different categories: A and B. The variable "Appropriateness" is a discrete quantitative variable assuming values from 0 to 10. "Age" of the respondents is a continuous quantitative variable, and "purchase intent" is an ordinal categorical variable assuming three different ordered categories.

In some contexts, you may find different terminology used to refer to similar data types. In quality control, categorical and discrete data are referred to as *attributes* and continuous data as *variables*.

When performing a statistical analysis, take into account the type of variable(s) you have, i.e. is it qualitative or quantitative? Different graphs, descriptive statistics, and inferential procedures must be used to study different types of data.

Stat Tool 1.2 Statistical Unit, Population, Sample

A *statistical unit* is the unit of observation (e.g. entity, person, object, product) for which data are collected. For each statistical unit, qualitative or quantitative variables are observed or measured.

The whole set of statistical units is the *population*. It may also be virtually infinite (e.g. all products of a production process).

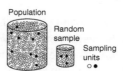

A *sample* is a subset of statistical units (sampling units) selected from the population in a suitable way. The sample size of a study is the total number of sampling units (see Figure 1.1).

Figure 1.1 Population, samples, sampling units.

When we use a sample to draw conclusions about a population, sample selection must be performed *at random*. Random sampling is carried out in such a way as to ensure that no element in the population is given preference over any other. Random sampling is used to avoid nonrepresentative samples of the population.

In *Example 1.1* the statistical unit is the single respondent.

Stat Tool 1.3 Descriptive and Inferential Analysis

Usually, the first step of a statistical analysis is *descriptive analysis*, where tables, graphs, and simple measures help to quickly assess and summarize important aspects of sample data.

When performing descriptive analysis, take into account the type of variable(s) present, i.e. is it qualitative (categorical) or quantitative? Different graphs, descriptive statistics, and inferential procedures have to be used to study different types of data.

The descriptive phase evaluates the following aspects:

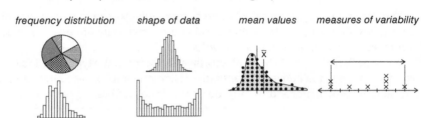

After outlining important sample data characteristics through descriptive statistics, the second step of a statistical analysis is *inferential analysis*, where sample findings are generalized to the referring population.

We often wish to answer questions about our processes or products to make improvements and predictions, save money and time, and increase customer satisfaction:

- What is the stability of a new formulation?
- What is the performance of a new product compared with the industry standard or products currently on the market?
- What is causing high levels of variation and waste during processing?

These questions are examples of *inferential problems*.
Inferential problems are usually related to:

Estimation of a population parameter (e.g. a mean)	⟶	What is the stability of a new formulation?
Comparison among groups	⟶	What is the performance of a new product compared with the industry standard or products currently on the market?
Assessing *relationships* among variables	⟶	What is causing high levels of variation and waste during processing?

Stat Tool 1.3 (Continued)

We can use several inferential techniques to answer different questions. Later on, we will review the following ones:

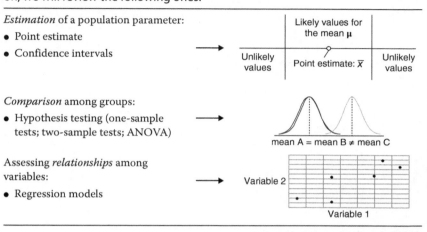

Estimation of a population parameter:
- Point estimate
- Confidence intervals

→

	Likely values for the mean μ	
Unlikely values	Point estimate: X̄	Unlikely values

Comparison among groups:
- Hypothesis testing (one-sample tests; two-sample tests; ANOVA)

→

mean A = mean B ≠ mean C

Assessing *relationships* among variables:
- Regression models

→ Variable 2

Variable 1

Stat Tool 1.4 Shapes of Data Distributions

Frequency distributions may be shown by tables or graphs. Use *bar charts* for categorical or quantitative discrete variables, *histograms* for continuous variables, and *dot plots* (especially useful for small data sets) for discrete or continuous variables.

By observing the frequency distribution of a categorical or quantitative variable, several *shapes* may be detected:

- When values or classes have similar percentages, the distribution is said to be fairly *uniform*. In a fairly uniform distribution there are no values or classes predominant over the others (a).
- When there is one value or class predominant over the others, the distribution is said to be *nonuniform* and *unimodal* with one peak (b).
- When there is more than one value or class predominant over the others, the distribution is said to be nonuniform and *multimodal* with more than one peak (c).

The value or class with the highest frequency is the *mode* of the distribution (see Figure 1.2).

Stat Tool 1.4 (Continued)

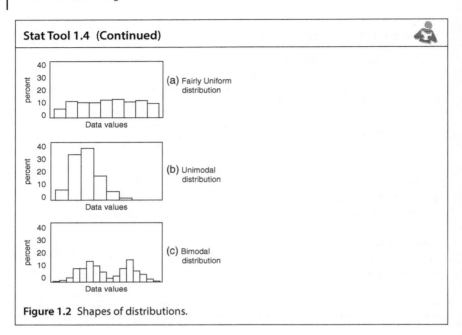

Figure 1.2 Shapes of distributions.

Stat Tool 1.5 Shapes of Data Distributions for Quantitative Variables

By observing the frequency distribution of a *quantitative* discrete or continuous variable, several *shapes* may be detected related also to the presence or absence of *symmetry* (Figures 1.3 and 1.4).

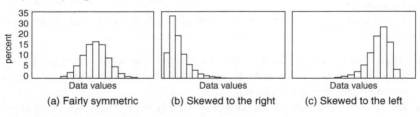

Figure 1.3 Shapes of distributions (symmetric and skewed distributions).

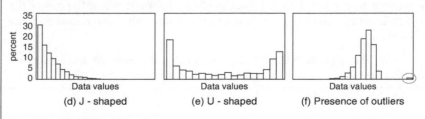

Figure 1.4 Other shapes of distributions.

Stat Tool 1.5 (Continued)

If one side of the histogram (or bar chart for quantitative discrete variables) is close to being a mirror image of the other, then the data are *fairly symmetric* (a). Middle values are more frequent, while low and high values are less frequent. If data are not symmetric, they may be *skewed to the right* (b) or *skewed to the left* (c). In (b) low and middle values are more frequent than high values. In (c) high and middle values are more frequent than low values.

If histograms (or bar charts for quantitative discrete variables) show ever-decreasing or ever-increasing frequencies, the distribution is said to be *J-shaped* (d). If frequencies are decreasing on the left side of the graph and increasing on the right side, the distribution is said to be *U-shaped* (e). Sometimes there are values that do not fall near any others. These extremely high or low values are called *outliers* (f).

Stat Tool 1.6 Measures of Central Tendency: Mean and Median

When quantitative data distributions tend to concentrate around certain values, we can try to locate these values by calculating the so-called measures of *central tendency*: the *mean* and the *median*. These measures describe the area of the distribution where most values occur.

The *mean* is the sum of all data divided by the number of data. It represents the "balance point" of a set of values.

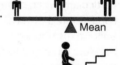

The *median* is the middle value in a *sorted* list of data. It divides data in half: 50% of data are greater than the median, 50% are less than the median.

For *symmetric* data, mean and median tend to be close in value (Figure 1.5):
In *skewed data* or *data with extreme values*, mean and median can be quite different. Usually for such data, the median tends to be a better indicator of the

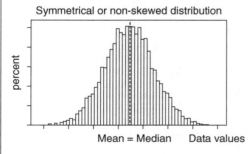

Figure 1.5 Mean and median in symmetric distributions.

Stat Tool 1.6 (Continued)

central tendency rather than the mean, because while the mean tends to be pulled in the direction of the skew, the median remains closer to the majority of the observations (Figure 1.6).

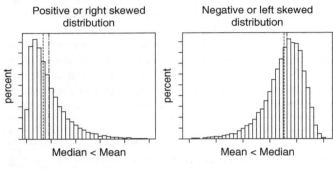

Figure 1.6 Mean and median in skewed distributions.

Stat Tool 1.7 Measures of Non-Central Tendency: Quartiles

Particularly when *numeric* data do not tend to concentrate around a unique central value (e.g. fairly uniform distributions), more than one descriptive measure is needed to summarize the data distribution. These measures are called *quantiles*.

The most common quantiles are *quartiles,* which are three values (first quartile Q_1, second quartile Q_2, and third quartile Q_3) corresponding to specific positions in the *sorted* list of data values (Figure 1.7).

75% of the data are less than Q_3 and 25% are greater than Q_3.
50% of the data are less than Q_2 and 50% are greater than Q_2.
25% of the data are less than Q_1 and 75% are greater than Q_1.

The first quartile is also known as the 25th percentile, the median as the 50th percentile, and the third quartile as the 75th percentile.

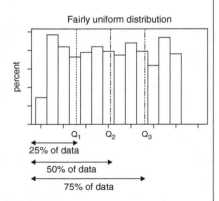

Figure 1.7 Quartiles.

Stat Tool 1.8 Measures of Variability: Range and Interquartile Range

Variability refers to how spread out a set of data values is.

Low variability   High variability

Consider the following graphs (see Figure 1.8):

- The two data distributions are quite different in terms of variability: the graph on the left shows more densely packed values (less variability), while the graph on the right reveals more spread out data (higher variability).
- The terms *variability*, *spread*, *variation*, and *dispersion* are synonyms, and refer to how spread out a distribution is.

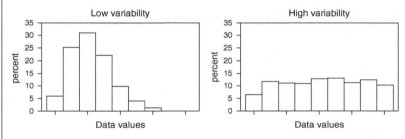

Figure 1.8 Frequency distributions and variability.

How can the spread of a set of *numeric* values be quantified?

The *range*, commonly represented as R, is a simple way to describe the spread of data values. It is the difference between the maximum value and the minimum value in a data set. The range can also be represented as the interval: (minimum value; maximum value).

A large range value (or a wide interval) indicates greater dispersion in the data. A small range value (or a narrow interval) indicates that there is less dispersion in the data.

Note that the range only uses two data values. For this reason, it is most useful in representing dispersion when data doesn't include outliers.

A second measure of variation is the *interquartile range*, commonly represented as IQR. It is the difference between the third quartile Q_3 and the first quartile Q_1 in a data set. IQR can also be represented as the interval: $(Q_1; Q_3)$. Fifty percent of the data are within this range: as the spread of these data increases, the IQR becomes larger.

The IQR is not affected by the presence of outliers.

Stat Tool 1.9 Measures of Variability: Variance and Standard Deviation

For *numeric* data, spread can also be measured by the *variance*. It accounts for *all the data* by measuring the *distance or difference between each value and the mean*. These differences are called *deviations*. The variance is the sum of squared deviations, divided by the number of values minus one. Roughly speaking, the variance (usually denoted by S^2) is the average of the squared deviations from the mean.

➤ *Example 1.2*. Suppose you observed the following numeric data with their dotplot (Figure 1.9):

8.1   8.2   7.6   9.0   7.5   6.9   8.1   9.0   8.3   8.1   8.2   7.6

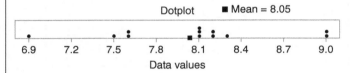

Figure 1.9 Dotplot.

The mean is equal to 8.05. Let's calculate the *deviations from the mean* and their squares:

Deviations	Squared deviations
8.1 − 8.05 = 0.05	$(0.05)^2 = 0.0025$
8.2 − 8.05 = 0.15	$(0.15)^2 = 0.0225$
7.6 − 8.05 = −0.45	$(−0.45)^2 = 0.2025$
9.0 − 8.05 = 0.95	$(0.95)^2 = 0.9025$
7.5 − 8.05 = −0.55	$(−0.55)^2 = 0.3025$
6.9 − 8.05 = −1.15	$(−1.15)^2 = 1.3225$
8.1 − 8.05 = 0.05	$(0.05)^2 = 0.0025$
9.0 − 8.05 = 0.95	$(0.95)^2 = 0.9025$
8.3 − 8.05 = 0.25	$(0.25)^2 = 0.0625$
8.1 − 8.05 = 0.05	$(0.05)^2 = 0.0025$
8.2 − 8.05 = 0.15	$(0.15)^2 = 0.0225$
7.6 − 8.05 = −0.45	$(−0.45)^2 = 0.2025$

Stat Tool 1.9 (Continued)

Now, by adding up the squared deviations to get the sum and dividing it by the number of values minus one, you obtain the variance:

$$S^2 = \frac{0.0025 + 0.0225 + \ldots + 0.2025}{11} = 0.359$$

The variance measures how spread out the data are around their mean. The greater the variance, the greater the spread in the data.

The variance is not in the same units as the data, but in squared units. If the data are in grams, the variance is expressed in squared grams, and so on. Thus, for descriptive purposes, its square root, called *standard deviation*, is used instead.

The *standard deviation* (usually denoted by S) quantifies variability *in the same units of measurement* as we measure our data.

Considering the previous example, the standard deviation is:

$$S = \sqrt{0.359} = 0.599$$

The greater the standard deviation, the greater the spread of data values around the mean.

Considering the mean and the standard deviation together and computing the range: mean $\pm$ S, we can say that data values vary on average from (mean $-$ S) to (mean $+$ S).

From the previous example the average range is:

$$\left(8.05 - 0.599;\ 8.05 + 0.599\right) = \left(7.451;\ 8.649\right)$$

The observed data vary on average from 7.5 to 8.6.

Stat Tool 1.10 Measures of Variability: Coefficient of Variation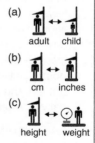

Another measure of variability for *numeric data* is the *coefficient of variation*.

It is calculated as follows:

$$CV\% = \left(\text{standard deviation}\,/\,\text{mean}\right) \times 100.$$

Being a dimensionless quantity, the coefficient of variation is a useful statistic for *comparing the spread* among several datasets, even if the means are different from one another (a), or data have different units (b), or data refer to different variables (c).

The higher the coefficient of variation, the higher the variability.

Stat Tool 1.11 Boxplots

So far, we have looked at three different aspects of numerical data analysis: *shape of the data, central and non-central tendency,* and *variability.*

Boxplots can be used to assess and compare these three aspects of *quantitative* data distributions, and to look for outliers.

Like histograms, boxplots work best with moderate to large sample sizes (at least 20 values).

Let's look at how a boxplot is constructed. It can be displayed horizontally or vertically:

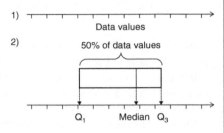

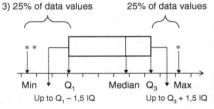

1) Start by drawing a horizontal or vertical axis in the units of the data values.
2) Draw a box to encompass 50% of middle data values. The left edge of the box is the first quartile Q_1. The right edge of the box is the third quartile Q_3. The width of the box is the interquartile range, IQR. Draw a line inside the box to denote the median.
3) Draw lines, called whiskers, on the left (to the minimum) and on the right (to the maximum) of the box to show the spread of the remaining data (25% of data points are below Q_1 and 25% are above Q_3). Several statistical softwares do not allow the whiskers to extend beyond one and a half times the interquartile range (1.5 × IQR). Any points outside of this range are outliers and are displayed individually by asterisks.

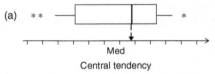

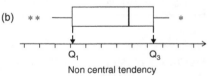

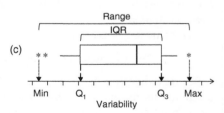

Boxplots help to summarize:
a) *Central tendency.* Look at the value of the median.
b) *Non-central tendency.* Look at the values of the first quartile Q_1 and the third quartile Q_3.

Stat Tool 1.11 (Continued)

c) *Variability*. Look at the length of the boxplot (range) and the width of the box (IQR).

d) *Shape of data*. Look at the position of the line of the median in the box and the position of the box between the two whiskers. In a symmetric distribution, the median is in the middle of the box and the two whiskers have the same length. In a skewed distribution, the median is closer to Q_1 (skewed to the right) or to Q_3 (skewed to the left) and the two whiskers do not have the same length (Figure 1.10).

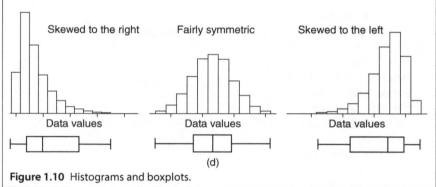

(d)

Figure 1.10 Histograms and boxplots.

Stat Tool 1.12 Basic Concepts of Statistical Inference

After describing important characteristics of sample data through descriptive statistics, the second step of a statistical analysis is usually *inferential analysis*, where sample findings are generalized to the referring population.

Inferential techniques use descriptive statistics such as:

sample mean "$\overline{x}$"   sample proportion "p"   sample standard deviation "S"

to draw conclusions about the corresponding unknown quantities of the population, called *parameters*:

population mean: "μ" population proportion "π" population standard deviation "σ"

Note that it is standard to use Greek letters for certain *parameters*, such as μ to stand for a population mean, σ for a population standard deviation, σ^2 for a population variance, and π for a proportion of statistical units having a characteristic of interest.

Stat Tool 1.12 (Continued)

A *statistic* (mean, proportion, variance) describes a characteristic of the sample (central tendency, variability, shape of data) and is *known*.

A *parameter* (mean, proportion, variance) describes a characteristic of the population (central tendency, variability, shape of data) and is *unknown*.

Statistical inference uses sample data to draw conclusions about a population with a known level of risk. In general, statistical inference proceeds as follows:

1) We are interested in a *population*.
2) We identify *parameters* of that population that will help us understand it better.
3) We take a *random sample* and compute *sample statistics*.
4) Through inferential techniques, we use the sample statistics to *infer* facts about the population parameters of interest.

POPULATION:
all customers of a product

PARAMETER:
Percentage of all customers highly satisfied

RANDOM SAMPLE of customers
STATISTIC: sample percentage
of customers highly satisfied

Stat Tool 1.13 Inferential Problems

As mentioned in Stat Tool 1.3, we often want to answer questions about our processes or products to make improvements and predictions, save money and time, and increase customer satisfaction:

- What is the stability of a new formulation?
- Which attributes of a product do consumers find most appealing?
- What is the performance of a new product compared with products currently on the market?
- What is causing high levels of variation and waste during processing?
- Can a process change reduce production time to get the product in stores more quickly?

Stat Tool 1.13 (Continued)

These questions are examples of *inferential problems*.
How can we use *inferential techniques* to answer these questions?
Inferential problems are usually related to:

Estimation of a population parameter:
- What is the stability of a new formulation?
- Which attributes of a product do consumers find most appealing?

Comparison of a population parameter to a specified value or among groups:
- What is the performance of a new product compared with the industry standard or products currently on the market?

Assessing *relationships* among variables:
- What is causing high levels of variation and waste during processing?
- Can a process change reduce production time to get the product in stores more quickly?

We may use several inferential techniques to answer different questions:

Estimation of a population parameter:
- Point estimate and confidence intervals

Comparison among groups:
- Hypothesis testing (one-sample tests; two-sample tests; analysis of variance, ANOVA)

Assessing *relationships* among variables:
- Regression models

**Stat Tool 1.14 Estimation of Population Parameters
and Confidence Intervals**

Let's introduce the problem of the estimation of a population parameter.

Because it is often impractical or impossible to gather data on the entire population, we must *estimate* the *population parameters* using *sample statistics*.

Statistics, such as the sample mean and standard deviation, are called *point estimators*.

A *point estimate* is a *single sample value* that approximates the true unknown value of a population parameter.

Stat Tool 1.14 (Continued)

- Point estimators:

  sample mean $\bar{x}$ sample proportion p   sample standard deviation S

- Population parameters:

population mean µ   population proportion π   population standard deviation σ

Point estimates, such as the sample mean or standard deviation, provide a lot of information, but they don't give us the full picture.

As it is highly unlikely that, for example, the sample mean and standard deviation we obtain are exactly the same as the population parameters, and to get a better sense of the true population values, we can use *confidence intervals*.

A *confidence interval* is a *range of likely values for a population parameter*, such as the population mean or standard deviation.

Usually, a confidence interval is a range:

$$(point\ estimate - \text{something};\ point\ estimate + \text{something})$$

Using confidence intervals, we can say that it is likely that the population parameter is somewhere within this range.

➤ *Example 1.3.* To illustrate this point, suppose that a research team wants to know the *mean* satisfaction score (from 0: completely not satisfied, to 10: completely satisfied) for the population of people who use a new formulation of a product.

Mean satisfaction score (population parameter) = ?

From a random sample of consumers, the *sample mean* is 6.8, and the *confidence interval* is CI = (6.2; 7.4).

Sample mean = 6.8
C.I. = (6.2; 7.4)

So the true unknown population mean satisfaction score is likely to be somewhere between 6.2 and 7.4.

The central point of the confidence interval is the sample mean: $\bar{x} = 6.8$ (point estimate of µ).

There's always a chance that the confidence interval won't contain the true population mean.

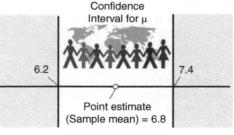

Confidence Interval for µ

6.2   7.4

Point estimate
(Sample mean) = 6.8

Stat Tool 1.14 (Continued)

When we use confidence intervals, we must decide how sure we need to be that the confidence interval contains the actual population parameter value, taking into account that we cannot be 100% sure.

We quantify how sure we need to be with a value called the *confidence level*, usually denoted by $(1 - \alpha)$.

The confidence level is set by the researcher before calculation of a confidence interval.

The most common confidence level is 95% (0.95). Other common levels are 90% and 99%.

The confidence level is how sure we are that the confidence interval contains the actual population parameter value.

➤ *Example 1.4.* To illustrate the meaning of the confidence level, let's return to the previous example and suppose we drew 100 samples from the same population and calculated the confidence interval for each sample.

If we used 95% confidence intervals, on average 95 out of 100 of the confidence intervals will contain the population parameter, while 5 out of 100 will not.

In practice when we calculate a 95% confidence interval for our sample, we are confident that our sample is one of the 95% samples for which CI covers the true parameter value.

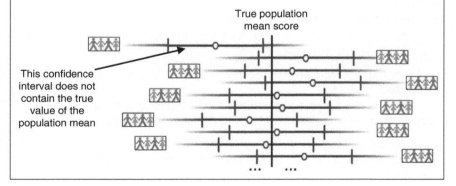

Stat Tool 1.15 Hypothesis Testing

A common task in statistical studies is the *comparison* of *mean values*, *variances*, *proportions*, and so on, to a hypothesized value of interest or among different groups, for example:

➤ What is the performance of a new product compared with the industry standard or products currently on the market?

To investigate any such questions, we can conduct an inferential procedure called a *hypothesis test*. *Hypothesis tests* allow us to make decisions on business problems based on statistically significant results, not based on intuition alone.

To begin with, the researchers need to determine which question they want to focus on and then define the *hypotheses*. A *statistical hypothesis* is a *claim about a population parameter* (e.g. about the mean or the standard deviation of a variable of interest).

Hypotheses should be based on our knowledge of the process, such as how a process has performed in the past or customers' expectations.

To perform a hypothesis test, we need to define the *null hypothesis* and the *alternative hypothesis*.

- *Null hypothesis* H_0: usually states that a population parameter, such as the population mean, *equals* a specified value or parameters from other populations.
  E.g. H_0: the mean performance of the new product is equal to the industry standard.
- *Alternative hypothesis* H_1: is the opposite of the null hypothesis, so it usually states that the population parameter *does not equal* a specified value or parameters from other populations.

  H_1: the mean performance of the new product is NOT equal to the industry standard.

  Sometimes the alternative hypothesis is *directional* or *one-sided*; that is, we suspect the population parameter is greater than or less than a given value.

- *Null hypothesis* H_0:
  E.g. H_0: the mean performance of the new product is equal to the industry standard.
- *One-sided alternative hypothesis* H_1:

  H_1: the mean performance of the new product is GREATER THAN the industry standard.
  Or
  H_1: the mean performance of the new product is LESS THAN the industry standard.

Stat Tool 1.15 (Continued)

After defining null and alternative hypotheses, the hypothesis test uses *sample data* to decide on *two possible conclusions*:

1) We can *reject the null hypothesis* in favor of the alternative hypothesis. If we reject the null hypothesis, we say that our result is *statistically significant*:
   E.g. H_0 (the mean performance of the new product is equal to the industry standard) is REJECTED.
2) We can *fail to reject the null hypothesis* and conclude that we do not have enough evidence to claim that the alternative hypothesis is true. We will say that our results are *NOT statistically significant*:
   E.g. H_0 (the mean performance of the new product is equal to the industry standard) is NOT REJECTED.

Because we are using sample data, decisions based on those hypothesis tests could be wrong.

Recall the *confidence level* $(1 - \alpha)$ that we discussed earlier (*Stat Tool 1.14*). The confidence level is how sure we are that the confidence interval contains the true population parameter value. This confidence level is set by the researcher and is usually equal to 95% (0.95) or 99% (0.99).

Let's consider the outcomes of a *hypothesis test*. If the null hypothesis is true and based on our sample data we fail to reject it, we make the *correct decision*, but if we reject it, we make an *error*. In hypothesis testing, the *probability of rejecting a null hypothesis that instead is true* is called *significance level* and is denoted by α. We always select it before performing the hypothesis test and it is usually equal to 5% (0.05) or 1% (0.01).

Confidence level and significance level are tools to quantify the *uncertainty* about our inferential conclusions.

Stat Tool 1.16 The p-Value

After establishing the null and alternative hypotheses and setting the significance level α, how do we decide to reject the null hypothesis?

When we conduct a hypothesis test, the results include a probability called *p-value*.

We use the *p-value* to determine whether we should *reject* or *fail to reject the null hypothesis*, by comparing it to the significance level α.

If the *p-value is less than* α, we *reject the null hypothesis* in favor of the alternative hypothesis (our result is *statistically significant*):
E.g. H_0 (the mean performance of the new product is equal to the industry standard) is REJECTED.

Stat Tool 1.16 (Continued)

If the *p-value is greater than or equal to* α, we *fail to reject the null hypothesis*. There is not enough evidence to claim that the alternative hypothesis is true (our results are *NOT statistically significant*):

E.g. H_0 (the mean performance of the new product is equal to the industry standard) is NOT REJECTED.

➢ *Example 1.5.* A researcher wants to investigate whether the performance of a new product differs with respect to the industry standard.

Suppose they set the significance level α equal to 0.05 (5%). The two hypotheses are:

H_0: The mean performance of the new product is equal to the industry standard.

H_1: The mean performance of the new product is NOT equal to the industry standard.

What if the p-value is 0.032? Would they reject or fail to reject the null hypothesis?

They should *reject the null hypothesis* H_0 because the p-value is less than the significance level α and conclude that the new product has a different mean performance with respect to the industry standard. The result of the test is *statistically significant* at $\alpha = 5\%$.

And, what if the p-value is 0.076? Would they reject or fail to reject the null hypothesis?

They should *fail to reject the null hypothesis* H_0 because the p-value is greater than the significance level α and conclude that there is not enough evidence to claim that the new product has a different mean performance with respect to the industry standard. The result of the test is *NOT statistically significant* at $\alpha = 5\%$.

The p-value indicates whether our results are statistically significant. However, just because our results are *statistically significant* doesn't mean that they are *practically significant*.

➢ *Example 1.6.* A production line manager attempts to reduce production time by modifying the process. They compare the mean production time of the old process with the mean production time of the new process using a hypothesis test.

The p-value is 0.032. Using an alpha of 0.05, would they reject or fail to reject the null hypothesis?

Stat Tool 1.16 (Continued)

Because the p-value is less than alpha, they reject the null hypothesis that the production times are equal. The difference between the mean production times is statistically significant.

Now, consider practical significance. If the difference between the two production times is five seconds, is that really practically significant? Is it worth the cost of implementing the process change?

Always consider the practical significance of your results and your knowledge of the process before reaching a final conclusion.

2

The Screening Phase

2.1 Introduction

In the initial stages of development of a new consumer product, a team is generally faced with many options and it is difficult to screen without the risk of missing opportunities to find preferred consumer solutions. When making crucial decisions regarding the next steps to take in product development, it is important to identify suitable statistical tools. Such tools make it possible to combine scientific expertise with data analytics capability and to support the understanding of each factor's relevance in providing an initial indication of possible synergistic effects among different factors. This can generally be achieved with relatively simple techniques, such as factorial designs, and analysis of the results through ANOVA (ANalysis Of VAriance).

This chapter is a guide for developers to properly organize a factorial experimental design through proper randomization, blocking, and replication (Anderson, M. J. and Whitcomb, P. J., 2015). It also provides suggestions on how to reduce variability and analyze data to achieve the best outcome.

A concrete example of product development is provided, looking at an air freshener and consumer preferences regarding constituent oils.

Specifically, this chapter deals with the following:

Topics	Stat tools
Introduction to DOE (Design Of Experiments) and guidelines for planning and conducting experiments	2.1, 2.6
Factors, levels, and responses	2.1
Screening experiments and factorial designs	2.2
Basic principles of factorial designs (randomization, blocking, replication)	2.3, 2.4, 2.5
ANOVA (ANalysis Of VAriance)	2.7
Model assumptions for ANOVA	2.8
Residual analysis	2.9

End-to-End Data Analytics for Product Development: A Practical Guide for Fast Consumer Goods Companies, Chemical Industry and Processing Tools Manufacturers, First Edition.
Rosa Arboretti, Mattia De Dominicis, Chris Jones, and Luigi Salmaso.
© 2020 John Wiley & Sons Ltd. Published 2020 by John Wiley & Sons Ltd.
Companion website: www.wiley.com/go/salmaso/data-analytics-for-pd

---

Learning Objectives and Outcomes

Upon completion of this chapter, you should be able to do the following:

Know the general context of DOE.
Distinguish between factors, levels, and responses.
Understand the meaning of randomization, blocking, and replication.
Know how to plan a screening experiment using a factorial design.
Know how to plan the statistical analysis of a screening experiment.
Apply and interpret the results of ANOVA.
Detect the significant effects (both main effects and interactions) on the responses.
Know how to check the assumptions of ANOVA through residual analysis.

2.2 Case Study: Air Freshener Project

An air freshener project aims to introduce a new scented oil to provide consistent and long-lasting fragrance to homes. Six oils have to be tested to determine if there are any consumer perceived differences among different combinations of the fragrances. Expert panelists will test different *combinations of oils* by assigning them a liking score varying from 1 (dislike extremely) to 5 (like extremely). We need to plan and analyze an experimental study in order to identify which oils affect consumers' preferences.

2.2.1 Plan of the Screening Experiment

During an experiment, investigators select some factors to systematically vary in order to determine their effect on a response variable (Stat Tool 2.1). For the present study:

- The *six oils* represent the *key factors.*
- The liking score represents the *response variable.*

Stat Tool 2.1 Experiments, Factors, Responses

Experiments are used to study the performance of a process so it can be systematically improved (Box, G. E. P., Hunter, J. S., and Hunter, W. G., 2005). Think about the process as a combination of conditions (procedures, materials, machines, and more) called *factors* that transforms an input into an output with desirable properties we can measure as *response variables*. Some of the process conditions are controllable by the experimenter (*controllable factors*), whereas others are uncontrollable (Figure 2.1).

Stat Tool 2.1 (Continued)

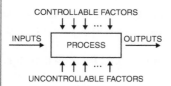

Figure 2.1 Process variables and conditions.

An experiment is generally a series of trials where the researcher intentionally varies one or more controllable factors, to observe changes in the output response and determine which factors affect the responses.

Factors and responses are statistical variables (Stat Tool 1.1), i.e. characteristics measured or controlled by the researcher on a sample of experimental units (Stat Tool 1.2). For controlled factors the experimenter chooses only a few *qualitative categories* or *numeric values*, called *levels*, to test in the experiment. Responses are *quantitative variables*, i.e. measurable characteristics on which factors may have an effect.

➤ *Example 2.1.* In a manufacturing process of plastic textile products, we control two factors on a sample of specimens (experimental units): type of additive (2 levels) and temperature (3 levels), and we measure tensile strength (response) to rate the fibers' ability to withstand stress (Figure 2.2).

FACTORS	FACTORS LEVELS		RESPONSE
TYPE OF ADDITIVE	Ⓐ Ⓑ		PLASTIC TENSILE STRENGTH
TEMPERATURE	HIGH (200°C) MEDIUM (150°C) LOW (100°C)		

Figure 2.2 Response, factors, and levels.

Suppose we know very little about consumers' preferences in relation to these oils and their combinations. At the beginning of an experimental study, if researchers have to investigate many factors, *screening experiments* based on factorial designs considering *k* factors, each at only two levels, are widely used to identify the key factors affecting responses. This also reduces the number of experimental conditions to test (also called *runs*). These designs are called *two-level factorial designs* (Stat Tool 2.2) and may be:

- *Full* (all combinations of the factors' levels are tested);
- *Fractional* (only a subset of all combinations is tested).

With fractional factorial designs, the experimenter can further reduce the number of runs to test, if they can reasonably assume that certain high-order interactions among factors are negligible.

In our example, as the number of potential input factors (the six oils) is large, we consider a two-level factorial design selecting the *two levels*: "*absent*" and "*present*" for each oil.

We will initially construct a full factorial design and then a fractional design in order to reduce the number of runs.

As two expert panelists will test the selected combinations of oils, assigning them a liking score from 1 to 5, panelists represent a blocking factor with two levels (Stat Tool 2.4): each *panelist* is a *block*.

Blocking is a technique for dealing with known and controllable nuisance factors, i.e. variables that probably have some effect on the response but are of no interest to the experimenter. Nuisance factors' levels are called *blocks*.

When feasible, the experiment should include at least two replicates of the design, i.e. the researcher should run each factor level combination at least twice. Replicates are multiple independent executions of the same experimental conditions (Stat Tool 2.5). As we have two blocks (the two panelists), this leads us to have two *replicates* of each run.

Stat Tool 2.2 DOE, Factorial Designs, Screening Experiments

DOE (Draper, N. R. and Pukelsheim F., 1996; Montgomery, D. C., 2008) and Statistical Analysis of the experimental data (Yandel, B. S., 1997) are two aspects of any experimental problem: the first refers to the planning of the experiment, the second to the statistical methods to be applied to the collected data to draw valid and objective conclusions.

When we have more than one factor, the correct approach is to plan a *factorial experiment* in which factors vary simultaneously, instead of one at a time. Factorial designs enable the researcher to investigate whether each factor has an effect on the response (*main effect*) and whether the factors interact (*interaction effect*), where an interaction exists between, for example, two factors if the effect of one factor on the response is not the same for all the levels of the other factor.

The experimental conditions, also called *treatments* or *runs*, tested on a sample of experimental units, are represented by:

- *Each level* of the factor, with only one factor;
- *Each combination of factor levels*, otherwise.

The entire set of runs is the *design*. See Figure 2.3 in reference to Example 2.1.

As the number of factors increases, the number of runs increases rapidly, making the experiment too expensive or time consuming to execute. However,

Stat Tool 2.2 (Continued)

FACTORS	FACTORS LEVELS	TREATMENT

TYPE OF ADDITIVE	A B	
TEMPERATURE	HIGH (200°C) MEDIUM (150°C) LOW (100°C)	LOW (100°C)

FULL FACTORIAL DESIGN

RUN	Additive	Temperature
1	A	LOW
2	A	MEDIUM
3	A	HIGH
4	B	LOW
5	B	MEDIUM
6	B	HIGH

FRACTIONAL FACTORIAL DESIGN

RUN	Additive	Temperature
1	A	LOW
3	A	HIGH
4	B	LOW
6	B	HIGH

Figure 2.3 Factors, levels, and treatments.

Figure 2.4 Full and fractional designs and treatments.

the researcher can decide to run all possible treatments by planning a *full factorial design* or only a suitable subset of the runs with a *fractional factorial design* (Figure 2.4).

In a full factorial design all main effects and interactions can be estimated, but with many factors and/or many levels, this may result in a prohibitive number of runs (Figure 2.5).

In a fractional factorial design some of the effects will be confounded (*aliased*), but the number of runs is reduced to a manageable size minimizing time and cost (Kounias, S. and Salmaso, L., 1998).

In the early stages of an experiment, when many factors are likely to be investigated, *screening experiments* with factorial designs considering k factors, each at only two levels, are widely used to identify the key factors affecting responses. These levels may be quantitative (e.g. two values of temperature) or qualitative (e.g. two machines, "high" and "low" levels of a factor, or the presence or absence of a factor). These are called *two-level factorial designs* and may be *full* (also referred to as 2^k factorial designs) or *fractional*. After detecting key factors, the experimenter can proceed with a *general factorial design* increasing the number of levels to optimize responses.

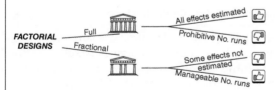

Figure 2.5 Advantages and disadvantages of full and fractional designs treatments.

Stat Tool 2.3 Basic Principles of Factorial Designs: Randomization

The three basic principles of experimental designs are *randomization*, *blocking*, and *replication*.

By *randomization* we mean that both the allocation of the experimental units to treatments and the order in which the individual runs are to be performed, are randomly determined. Computer software programs help the researcher to plan and construct a factorial design and generally present the runs in random order by default (Figure 2.6).

FACTORIAL DESIGN			RANDOMIZED FACTORIAL DESIGN		
RUN	ADDITIVE	TEMPERATURE	RUN	ADDITIVE	TEMPERATURE
1	A	LOW	5	B	MEDIUM
2	A	MEDIUM	1	A	LOW
3	A	HIGH	2	A	MEDIUM
4	B	LOW	4	B	LOW
5	B	MEDIUM	6	B	HIGH
6	B	HIGH	3	A	HIGH

Figure 2.6 Randomized design for plastic tensile strength (Example 2.1).

Many statistical methods of data analysis require response measurements (also called *observations*) to be *independent*. Randomization usually makes this assumption valid.

Randomization also helps in averaging or balancing out effects of *extraneous factors* (noise factors), that could have an effect on the response and that are difficult to be measured by the experimenter or are not known (Figure 2.7).

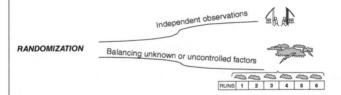

Figure 2.7 Advantages of randomization.

Stat Tool 2.4 Basic Principles of Factorial Designs: Blocking

Blocking is a technique for dealing with known and controllable *nuisance factors*, blocking out their potential effect on the response. A nuisance factor is a factor that probably has some effect on the response, but it's of no interest to the experimenter (e.g. time, batches of material, operators, machines). However, as it can add variability in the response, it needs to be taken into account, and improve the precision in the analysis (Figure 2.8).

Typically, each level of the nuisance factor becomes a block. If a complete replicate of the basic experiment is conducted in each block, the experiment is called *randomized complete block design* (RCBD) (Figure 2.9).

Stat Tool 2.4 (Continued)

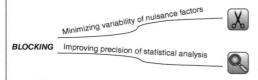

Figure 2.8 Advantages of blocking.

BASIC EXPERIMENT: FULL FACTORIAL DESIGN			TENSILE TESTING MACHINE (BLOCKING FACTOR)					
			MACHINE 1			MACHINE 2		
RUN	ADDITIVE	TEMPERATURE	RUN	ADDITIVE	TEMPERATURE	RUN	ADDITIVE	TEMPERATURE
1	A	LOW	5	B	MEDIUM	1	A	LOW
2	A	MEDIUM	2	A	MEDIUM	3	A	HIGH
3	A	HIGH	4	B	LOW	2	A	MEDIUM
4	B	LOW	1	A	LOW	6	B	HIGH
5	B	MEDIUM	6	B	HIGH	4	B	LOW
6	B	HIGH	3	A	HIGH	5	B	MEDIUM

Figure 2.9 RCBD for plastic tensile strength.

Stat Tool 2.5 Basic Principles of Factorial Designs: Replication

By replication, or replicates, we mean that we assign each factor combination to more than one experimental unit. In other words, replicates are multiple independent executions of the same experimental conditions. They are considered separate runs, processed individually in the experiment, and should be run in random order.

➢ *Example 2.1*. Plastic tensile strength.

For example, runs 9 and 3 are two replicates of the combination "additive A" and "High temperature" (Figure 2.10).

Replication allows the experimenter to obtain an estimate of the *experimental error* required in the statistical analysis of data in determining whether observed differences are really statistically different. It also allows us to calculate a *more precise estimate* of the true mean response for factors' levels (Figure 2.11).

Note that a replicate is not a *repeated* or *duplicate measurement* of the response. *Repeated measurements* often reflect the inherent variability in the measurement system.

RUNS	ADDITIVE	TEMPERATURE
9	A	HIGH
1	A	LOW
3	A	HIGH
7	A	LOW
5	B	MEDIUM
10	B	LOW
6	B	HIGH
2	A	MEDIUM
8	A	MEDIUM
12	B	HIGH
11	B	MEDIUM
4	B	LOW

Figure 2.10 Presence of replications for plastic tensile strength (Example 2.1).

Stat Tool 2.5 (Continued)

➤ *Example 2.2.* The cleaning performance (response) of three formulations of a stain removal product (factor) is measured at four different positions on

Estimate of experimental error

REPLICATION More precise estimates of mean response

Figure 2.11 Advantages of replication.

each piece of a sample of fabrics. The individual measurements for each position on a piece of fabric are not replicates but repeated measurements, because the four positions were processed together in the experiment, receiving the same treatment simultaneously. With repeated measurements, their *average* is usually the correct response variable to analyze.

To create the design for our screening experiment, let's proceed in the following way:

Step 1 – Create a full factorial design.
Step 2 – Alternatively, choose the desired fractional design.
Step 3 – Assign the designed combinations of factor levels to the experimental units and collect data for the response variable.

2.2.1.1 Step 1 – Create a Full Factorial Design

Let's begin to create a full factorial design, considering two levels (absent/present) for each factor and a blocking factor with two blocks (the two panelists).

To create a full factorial design, go to:

Stat > DOE > Factorial > Create Factorial Design

Choose **2-level factorial (default generators)** and in **Number of factors,** specify 6. Click on **Designs**. In the next dialog box, select **Full Factorial.** You can see that this design will include 64 factor-level combinations. In **Number of replicates for corner points** and in **Number of blocks,** specify 2. Click **OK,** and in the main dialog box select **Factors**. Enter the names of the factors (Oil1 to Oil6); select **Text** in **Type** as they are categorical variables, and specify **NO** and **YES** in **Low** and **High** to represent respectively the "absence" and the "presence" of each factor.

Proceed by clicking **Options** in the main dialog box and check the option **Randomize runs,** so that both the allocation of the experimental units and the order of the individual runs will be randomly performed (Stat Tool 2.3).

Clicking **OK** in the main dialog box, Minitab shows the design of the experiment in the worksheet (only the first five runs are shown here as an example), adding some extra columns that are useful for the statistical analysis, and some information on the design in the Session window.

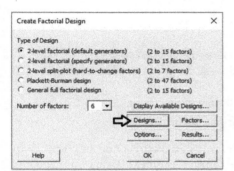

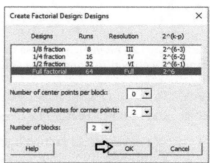

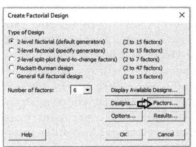

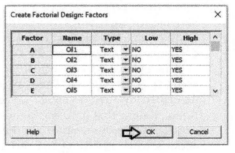

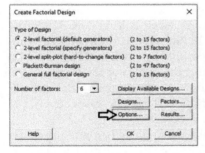

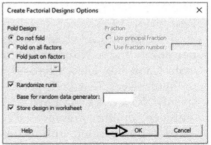

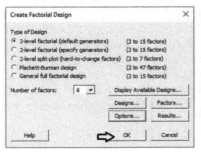

↓	C1	C2	C3	C4	C5-T	C6-T	C7-T	C8-T	C9-T	C10-T
	StdOrder	RunOrder	CenterPt	Blocks	Oil1	Oil2	Oil3	Oil4	Oil5	Oil6
1	83	1	1	2	NO	YES	NO	NO	YES	NO
2	119	2	1	2	NO	YES	YES	NO	YES	YES
3	81	3	1	2	NO	NO	NO	NO	YES	NO
4	108	4	1	2	YES	YES	NO	YES	NO	YES
5	97	5	1	2	NO	NO	NO	NO	NO	YES

2.2.1.2 Step 2 – Alternately, Choose the Desired Fractional Design

Suppose we need to construct a fractional design to reduce the number of runs to test.

 To create a fractional factorial design, go to:

Stat > DOE > Factorial > Create factorial Design

Choose **2-level factorial (default generators)** and in **Number of factors** specify 6. Click on **Designs**. In the next dialog box, you can select three types of fractional designs, respectively, with 8, 16, and 32 runs for each block. These designs correspond to specific fractions of the full design (e.g. 1/8, 1/4, or 1/2 fraction). Fractional designs can't estimate all the main effects and interactions among factors separately from each other: some effects will be confused (aliased) to other effects. The way in which the effects are aliased is described by the so-called "Design Resolution." By choosing fractional designs that have the highest resolution, the experimenter can usually obtain information on the main effects and low-order interactions, while assuming that certain high-order interactions are negligible. In our case, let's select, for example, a **1/2 fraction design** of **Resolution VI**. You can see that this design will include 32 factor-level combinations for each block. In **Number of replicates for corner points** and in **Number of blocks** specify 2. Click **OK,** and in the main dialog box select **Factors**. Enter the names of the factors (Oil1 to Oil6); select **Text** in **Type,** as they are categorical variables, and specify **NO** and **YES** in **Low** and **High** to represent, respectively, the "absence" and the "presence" of each factor.

Proceed by clicking **Options** in the main dialog box and check the option **Randomize runs** as before. Clicking **OK** in the main dialog box, Minitab shows the design in a new worksheet and more information is in the session panel. Note that for the fractional design, in the Session window the alias structure lists the aliasing between main effects and interaction. For example, A + BCDEF means that factor A (Oil1) is aliased with the four-order interaction BCDEF; we will estimate the effect of (A + BCDEF), but if we can reasonably assume that the interaction BCDEF is negligible, we will assign the total estimate of the effect to the main effect A.

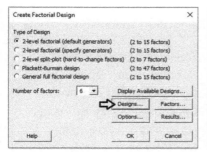

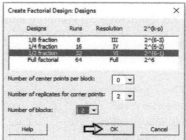

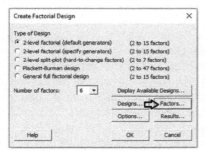

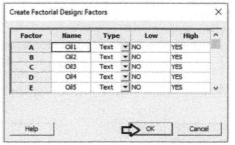

2.2.1.3 Step 3 – Assign the Designed Factor Level Combinations to the Experimental Units and Collect Data for the Response Variable

For each block, collect the data for the response variable following the order provided by the column **RunOrder** in the full or fractional design. For example, in the full factorial design created in step 1, the second panelist (Blocks = 2) will first test the fragrance where Oil2 and Oil5 are present, while Oil1, Oil3, Oil4, and Oil6 are absent. The two panelists will test each run by assigning it a liking score, varying from 1 (dislike extremely) to 5 (like extremely). Once all the designed factor-level combinations have been tested, enter the recorded response values (liking scores) into the worksheet containing the design. The worksheet will be like the one below (only the first five records are shown here as an example for the fractional design obtained in step 2). Now you are ready to proceed with the statistical analysis of the collected data, but before doing this, have a look at Stat Tool 2.6 to get a summary overview of the main aspects of the design and analysis of experiments.

	C1	C2	C3	C4	C5-T	C6-T	C7-T	C8-T	C9-T	C10-T	C11
↓	StdOrder	RunOrder	CenterPt	Blocks	Oil1	Oil2	Oil3	Oil4	Oil5	Oil6	Response
1	38	1	1	2	YES	NO	YES	NO	NO	NO	2.5
2	61	2	1	2	NO	NO	YES	YES	YES	YES	4.0
3	41	3	1	2	NO	NO	NO	YES	NO	YES	4.8
4	48	4	1	2	YES	YES	YES	YES	NO	NO	4.7
5	62	5	1	2	YES	NO	YES	YES	YES	NO	2.1

Worksheet 2 ***

Stat Tool 2.6 Guidelines for Planning and Conducting Experiments

In *Design and Analysis of Experiments* (Montgomery, D. C., 2017), the author summarizes a series of steps to follow for a correct statistical approach in designing and analyzing an experiment.

Designing experiments
{
Statement of the problem
Selection of the response variable
Choice of factors and levels
Choice of experimental design
Performing the experiment
Statistical analysis of data
Conclusions and recommendatior
}

The following diagrams outline some relevant key points.

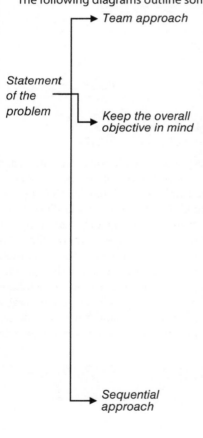

Statement of the problem
→ *Team approach*
→ *Keep the overall objective in mind*
→ *Sequential approach*

Develop a clear and generally accepted statement of the problem soliciting input from all concerned parties (engineering, chemists, manufacturing, marketing, management, customers, operating personnel, …).

- *Characterization or factor screening.* In a new process or system, the number of potential input variables (factors) is large. We need to *reduce* them by *identifying the key input factors* that affect response variable(s).
- *Optimization* of a mature or reasonably *well-understood system* that has been previously characterized.
- *Confirmation.* Is the system *performing the same way* now as it did in the past?
- *Discovery.* What happens if we *explore new operating conditions*, materials, variables, etc.?
- *Stability or robustness.* How can we *reduce variability in the response variable* arising from uncontrolled sources?

Employ a *series of smaller experiments*, each with a specific objective, rather than a single comprehensive experiment.

Stat Tool 2.6 (Continued)

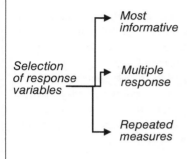

| | Most informative | Choose the variable that really provides *useful information* about the process under study. |

Selection of response variables

Multiple response — Multiple responses representing *multiple aspects of the process output* may be of interest.

Repeated measures — In situations where gauge capability is poor or measurement error is large, the experimenter *measures the response on each experimental unit several times* and uses the *average* of these repeated measures as the observed response.

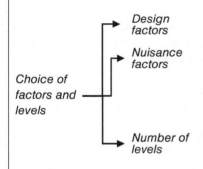

Choice of factors and levels

Design factors — *Key Factors*: Factors of interest.

Nuisance factors — *Factors not of interest*: They may have effects on the response that must be accounted for through the blocking technique, randomization, and analysis of covariance.

Number of levels — *Two-level full* and *fractional factorial designs* (factors have only two levels) are often used in factor screening or process characterization to "screen" for the really important factors that influence process output responses. *General full factorial designs* (factors can have any number of levels) may be used in optimization/confirmatory experiments.

Choice of experimental design

Sample size
Number of replicates
Selection of a suitable run order for trials
Blocking

Statistical software packages and MINITAB® support this phase of the experimental design (Mathews, P. G., 2005; Greenfield, T. and Metcalfe, A., 2007), proposing a selection of designs for consideration and providing worksheets with the order of the randomized runs.

Stat Tool 2.6 (Continued)

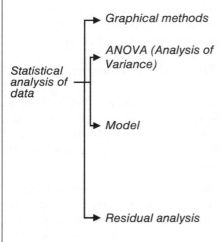

Performing the experiment

→ *Monitoring*

It is important to *monitor the process carefully* to ensure that everything is being done according to plan and to avoid errors in experimental procedures.

→ *Pilot runs*

When possible, *a few pilot runs* prior to conducting the experiment may provide information about the overall experimental design and on consistency of experimental material, measurement system, and experimental error.

Statistical analysis of data

→ *Graphical methods*

They play an important role in data analysis and interpretation.

→ *ANOVA (Analysis of Variance)*

Many of the questions that the experimenter wants to answer can be solved by *hypothesis testing* and *confidence interval* estimation procedures.

→ *Model*

It is also helpful to present the results in terms of a *model*, i.e. an equation derived from the data that expresses the *relationship* between the response and the important design factors.

→ *Residual analysis*

Residual analysis and model *adequacy checking* are also important analysis techniques.

Conclusions and recommendations

→ *Practical conclusions*

Once the data have been analyzed, the experimenter draws *practical conclusions* and recommends a *course of action*.

Recognize the *difference between practical and statistical significance*. A statistically significant difference may be too small to be of practical value.

→ *And ...*

Stat Tool 2.6 (Continued)

Experiments are usually iterative The *learning process* advances by formulating hypotheses, performing experiments to investigate them, formulating new ones on the basis of the results, and so on.

It is usually a mistake to design a single, large, comprehensive experiment at the start of the study.

A *sequential approach* employing a series of smaller experiments is usually a better strategy.

Use your *nonstatistical knowledge of the problem* in choosing factors and levels, number of replicates, interpreting the results, and so forth.

Using a designed experiment is no substitute for *thinking about the problem*.

Do the *pre-experiment planning carefully* and select a *reasonable design*.

Keep the *design and analysis as simple as possible*.

Do not be overzealous in the use of complex, sophisticated statistical techniques.

Relatively *simple design and analysis methods* are almost always the best.

Conclusions and recommendations

2.2.2 Plan of the Statistical Analyses

The main interest in the statistical analysis of a screening experiment is to detect which factors (if any) show a significant contribution to the explanation of the response variable and also something about how the factors interact. For each factor, it is possible to evaluate if the mean response varies between the two factor levels, taking into account the effect of other factors on the response. The appropriate procedure to use for this purpose is the ANOVA (Stat Tool 2.7).

Let's consider the fractional design obtained in step 2 (Section 2.2.1) and suppose to have collected the response data (File: Air_freshener_Project.xlsx). Remember that if the experimenter can reasonably assume that certain high-order interactions are negligible, a fractional design can give useful information on the main effects and low-order interactions. The factors identified as important through the statistical analysis of collected data will then be investigated more thoroughly in subsequent experiments.

To analyze our screening experiment, let's proceed in the following way:

Step 1 – Perform a descriptive analysis (Stat Tool 1.3) of the response variable.

Step 2 – Apply the ANOVA to estimate the effects and determine the significant ones.

Step 3 – If required, reduce the model to include the significant terms.

The variables setting is the following:

- Columns from C1 to C3 are related to the previous creation of the fractional factorial design.
- Column C4 is the blocking factor, assuming values 1 and 2.
- Variables OIL1–OIL6 are the categorical factors, each assuming two levels.
- Variable "Response" is the *response variable*, assuming values from 1 to 5.

Column	Variable	Type of data	Label
C4	Block	Numeric data	Blocking factor representing the two panelists 1 and 2
C5–C10	OIL1–OIL6	Categorical data	The six oils for fragrances, each with levels: NO, YES
C11	Response	Numeric data	The liking score, varying from 1 to 5

Air_freshener_Project.xlsx

2.2.2.1   Step 1 – Perform a Descriptive Analysis of the Response Variable

As the response is a quantitative variable, use a dotplot (recommended for small samples with less than 50 observations) or a histogram (for moderate or large datasets with a sample size greater than 20) to describe how the liking scores occurred in our sample. Add a boxplot (for moderate or large datasets) and calculate means and measures of variability to complete the descriptive analysis of the response.

 To display the histogram, go to: **Graph** > **Histogram**

Check the graphical option **Simple**, and in the next dialog box select "Response" in **Graph variables**. Then, select the option **Scale** and in the next screen **Y-Scale Type** to display percentages in the histogram. Then click **OK** in the main dialog box.

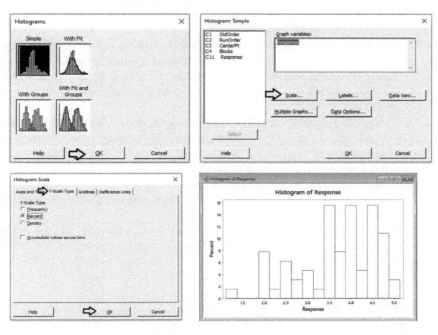

To change bar width and number of intervals, double-click on any bar and select the **Binning** tab. A dialog box will open with several options to define. Specify 5 in **Number of intervals** and click **OK**.

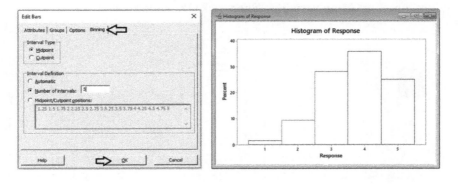

Use the following tips to change the appearance of the histogram:

- To change the scale of the horizontal axis: double-click on any scale value on the horizontal axis. A dialog box will open with several options to define.
- To change the scale of the vertical axis: double-click on any scale value on the vertical axis. A dialog box will open with several options to define.
- To change the width and number of intervals of the bars: double-click on any bar and then select the Binning tab. A dialog box will open with several options to define.
- To change the graph window (background and border): double- click on the area outside the border of the histogram. A dialog box will open with several options to define.
- To add reference lines: right-click on the area inside the border of the box-plot, then select Add > Data display > Mean symbol.
- To change the title, labels, etc.: double-click on the object to change. A dialog box will open with several options to define.
- To insert a text box, lines, points, etc.: from the main menu at the top of the session window, go to Tools > Toolbars > Graph Annotation Tools the Session window.

To display the boxplot, go to: **Graph** > **Boxplot**

Check the graphical option **Simple** and in the next dialog box, **select** "Response" in **Graph variables**. Then click **OK** in the main dialog box.

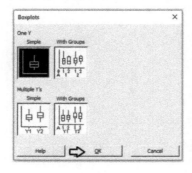
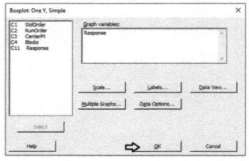

To change the appearance of a boxplot, follow the tips provided for histograms, as well as the following:

- To display the boxplot horizontally: double-click on any scale value on the horizontal axis. A dialog box will open with several options to define. Select from these options: Transpose value and category scales.
- To add the mean value to the boxplot: right-click on the area inside the border of the boxplot, then select Add > Data display > Mean symbol.

To display the descriptive measures (means, measures of variability, etc.), go to:

Stat > Basic Statistics > Display Descriptive Statistics

Select "Response" in **Variables**, then click on **Statistics** to open a dialog box displaying a range of possible statistics to choose from. Leave the default options and select **Interquartile range** and **Range**.

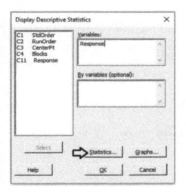
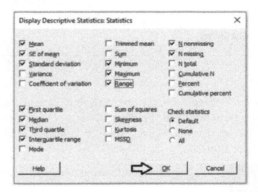

2.2.2.1.1 *Interpret the Results of Step 1* Let's describe the shape, central, and non-central tendency and variability of the distribution of the liking scores, presented in the histogram and boxplot below.

Shape: The distribution of the liking scores is skewed to the left: middle and high scores are more frequent than low scores (Stat Tools 1.4–1.5, 1.11).

Central tendency: The central tendency (the median score) is equal to 3.8: about 50% greater than or equal to 3.8. The mean is equal to 3.69. Mean and median are quite close in value (Stat Tools 1.6, 1.11).

Non-central tendency: 25% of the scores are less than 3.0 (first quartile Q_1) and 25% are greater than 4.5 (third quartile Q_3) (Stat Tools 1.7, 1.11).

Variability: Observing the length of the boxplot (range) and the width of the box (interquartile range, IQR), the distribution shows high variability. Its scores vary from 1.3 (minimum) to 4.5 (maximum). 50% of middle evaluations vary from 3.0 to 4.5 (IQR = maximum observed difference between evaluations equal to 1.5) (Stat Tools 1.8, 1.11).

The average distance of the scores from the mean (standard deviation) is 0.89: Scores vary on average from (3.69−0.89) = 2.80 to (3.69 + 0.89) = 4.58 (Stat Tool 1.9).

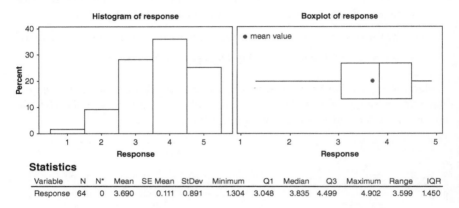

Statistics

Variable	N	N*	Mean	SE Mean	StDev	Minimum	Q1	Median	Q3	Maximum	Range	IQR
Response	64	0	3.690	0.111	0.891	1.304	3.048	3.835	4.499	4.902	3.599	1.450

2.2.2.2 Step 2 – Apply the Analysis of Variance to Estimate the Effects and Determine the Significant Ones

The appropriate procedure to analyze a factorial design is the ANOVA (Stat Tool 2.7).

 To analyze the screening design, go to:

Stat > DOE > Factorial > Analyze Factorial Design

Select the numeric response variable **Response** and click on **Terms**. At the top of the next dialog box, in **Include terms in the model up through order**, specify 2 to study all main effects and the two-way interactions. Select **Include blocks in the model** to consider blocks in the analysis. Click **OK** and in the main dialog box, choose **Graphs**. In **Effects plots**, choose **Pareto chart** and **Normal Plot**. These graphs help you identify the terms (factors and/or interactions) that influence the response and compare the relative magnitude of the effects, along with their statistical significance. Click **OK** in the main dialog box and Minitab shows the results of the analysis both in the session window and through the required graphs.

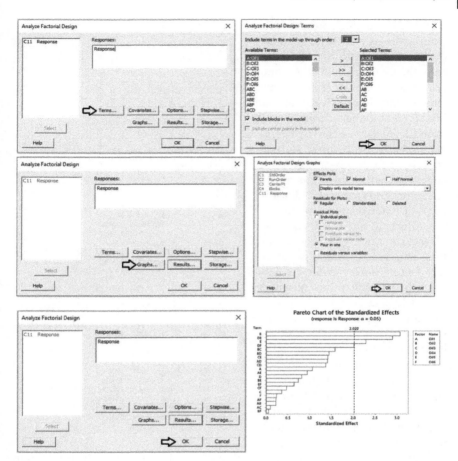

2.2.2.2.1 Interpret the Results of Step 2
Let us first consider the **Pareto chart** that shows which terms contribute the most to explain the response. Any bar extending beyond the reference line (considering a significance level equal to 5%) is related to a significant effect.

In our example, the main effects for Oil2 and Oil5 are statistically significant at 0.05. Furthermore, the interaction between Oil4 and Oil5 is marked as statistically significant, while the interaction between Oil4 and Oil6 is nearly significant. From the Pareto chart, you can detect which effects are statistically significant but you have no information on how these effects affect the response.

Use the **normal plot of the standardized effects** to evaluate the direction of the effects. In this case, the line represents the situation in which all the effects (main effects and interactions) are 0. Effects that depart from 0 and from the line are statistically significant. Minitab shows statistically significant and nonsignificant effects by giving the points different colors and shapes. In addition, the plot indicates the direction of the effect. Positive effects (displayed on the right side of the line) increase the response when the factor moves from its low value to its high value. Negative effects (displayed on the left side of the graph) decrease the response when moving from the low value to the high value.

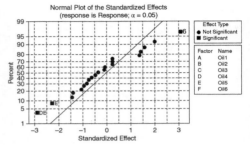

In our example, the main effects for Oil2 and Oil5 are statistically significant. The low and high levels represent the absence and presence of the specific oil; therefore, the liking scores seem to increase when Oil2 is present and Oil5 is absent. Furthermore, the interaction between Oil4 and Oil5 is marked as statistically significant.

Analysis of Variance

Source	DF	Adj SS	Adj MS	F-Value	P-Value
Model	22	25.2442	1.14746	1.90	0.037
Blocks	1	0.0886	0.08861	0.15	0.704
Linear	6	10.0254	1.67090	2.77	0.024
Oil1	1	0.6781	0.67810	1.12	0.295
Oil2	1	5.6870	5.68696	9.42	0.004
Oil3	1	0.0770	0.07703	0.13	0.723
Oil4	1	0.4029	0.40288	0.67	0.419
Oil5	1	3.1466	3.14660	5.21	0.028
Oil6	1	0.0339	0.03386	0.06	0.814
2-Way Interactions	15	15.1302	1.00868	1.67	0.097
Oil1*Oil2	1	0.0302	0.03022	0.05	0.824
Oil1*Oil3	1	0.0056	0.00561	0.01	0.924
Oil1*Oil4	1	1.1990	1.19897	1.99	0.166
Oil1*Oil5	1	0.5412	0.54121	0.90	0.349
Oil1*Oil6	1	0.0304	0.03035	0.05	0.824
Oil2*Oil3	1	1.5012	1.50119	2.49	0.122
Oil2*Oil4	1	1.2557	1.25567	2.08	0.157
Oil2*Oil5	1	0.3336	0.33363	0.55	0.461
Oil2*Oil6	1	0.0023	0.00231	0.00	0.951
Oil3*Oil4	1	1.1294	1.12945	1.87	0.179
Oil3*Oil5	1	1.2056	1.20563	2.00	0.165
Oil3*Oil6	1	0.1327	0.13272	0.22	0.642
Oil4*Oil5	1	5.0718	5.07176	8.40	0.006
Oil4*Oil6	1	2.4467	2.44668	4.05	0.051
Oil5*Oil6	1	0.2448	0.24477	0.41	0.528
Error	41	24.7437	0.60351		
Total	63	49.9879			

In the ANOVA table we examine p-values to determine whether any factors or interactions, or even the blocks, are statistically significant. Remember that the p-value is a probability that measures the evidence against the null hypothesis. Lower probabilities provide stronger evidence against the null hypothesis. For the main effects, the null hypothesis is that there is not a significant difference in the mean liking score across each factor's low level (absence) and high level (presence). For the two-factor interactions, H_0 states that the relationship between a factor and the response does not depend on the other factor in the term. With respect to the blocks, the null hypothesis is that the two panelists do not change the response. Usually we consider a significance level alpha equal to 0.05, but in an exploratory phase of the analysis we may also consider a significance level of 0.10.

When the p-value is greater than or equal to alpha, we fail to reject the null hypothesis. When it is less than alpha, we reject the null hypothesis and claim statistical significance.

So, setting the significance level $\alpha = 0.05$, which terms in the model are significant in our example? The answer is: Oil2, Oil5 and the interaction between Oil4 and Oil5. Now you may want to reduce the model to include only significant terms.

When using statistical significance to decide which terms to keep in a model, it is usually advisable not to remove entire groups of terms at the same time. The statistical significance of individual terms can change because of the terms in the model. To reduce your model, you can use an automatic selection procedure – the stepwise strategy – to identify a useful subset of terms, choosing one of the three commonly used alternatives (standard stepwise, forward selection, and backward elimination).

Stat Tool 2.7 ANOVA, Analysis Of VAriance

A common task in statistical studies is to compare the mean of a quantitative variable of interest among more than two groups. Use ANOVA to determine whether *means of more than two groups differ*. Usually, ANOVA is applied in an experimental context. ANOVA requires the following:

- A *response* (quantitative variable) is measured on sample units.
- One or more *factors* (categorical variables or quantitative variables) are controlled by the experimenter and used to identify more than two groups (treatments), related to factor levels or combinations of factor levels.

In other words, for each factor, two or more values (*levels*) are set by the experimenter. These levels or combinations of levels (when considering more than one factor) are called *treatments* and define *groups* across which the *means of a response* are compared.

For example, a researcher is interested in comparing the mean time of dissolution for three products formulated with three different doses of an additive. Here, the response is the time of dissolution and the factor is the additive dose.

With ANOVA you are studying the relationship between a response variable and one or more controlled factors.

When groups are defined on the basis of a *single factor*, the ANOVA is called one-way ANOVA. When groups are defined on the basis of *more than one factor*, the ANOVA is called multi-way or multi-factor ANOVA.

In ANOVA, the F test is applied to determine whether the response means differ among treatments. The null hypothesis H_0 states that all the means are equal. The alternative hypothesis H_1 states that the means are not all equal. If the *p-value* of the test is less than the significance level alpha, *reject the null hypothesis* of equality of means (Stat Tools 1.15, 1.16).

If you reject the null hypothesis, then apply a *multiple comparisons procedure* to assess how the means differ from one another. One of these procedures is *Tukey's multiple comparison procedure*, which considers all pairwise comparisons between treatments.

Stat Tool 2.7 (Continued)

The ANOVA procedure also presents the results in terms of a *model*, i.e. an equation derived from the data that expresses the *relationship* between the response and the factors. You can use this model to predict the response for each treatment. These *predicted* or *fitted values* are simply the *sample averages* of the observed response data for each treatment.

Stat Tool 2.8 Model assumptions for ANOVA

The use of ANOVA to formally test for no differences in treatment means requires that certain assumptions be satisfied:

Independent measurements	Each measure of the response variable is taken independently of the other measures.
Normality	For each group, response values are assumed to come from a normal distribution.
Homoscedasticity	There is similar variability of responses across groups.

➤ *Example 2.3.* A chemist wants to compare the technical performance of three different formulations of a stain-removal product (factor). Performance (response) is measured as the reflectance (after wash) of a fabric soiled with a specific stain.

Each formulation of the stain-removal product is *randomly* applied to a sample of pieces (replicates) of a fabric soiled with a specific stain. Performance is then measured on each piece of fabric, such that each measure is taken *independently* of the other measures.

For each formulation, experimental data is assumed to come from a *normal distribution*.

Reflectance should display *similar variability across groups* (formulations).

We can check these assumptions by performing a *residual analysis* (Figure 2.12).

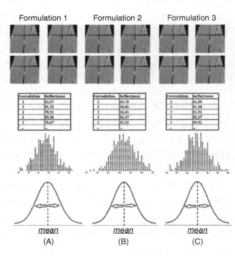

Figure 2.12 Model assumptions for ANOVA (Example 2.4).

2.2.2.3   Step 3 – If Required, Reduce the Model to Include the Significant Terms

To reduce the model, go to:

Stat > DOE > Factorial > Analyze Factorial Design

Select the numeric response variable **Response** and click on **Terms**. In **Include terms in the model up through order,** specify 2 to study all main effects and the two-way interactions. Select **Include blocks in the model** to consider blocks in the analysis. Click **OK** and in the main dialog box, choose **Stepwise**. In **Method** select **Backward elimination** and in **Alpha to remove** specify **0.05**, then click **OK**. In the main dialog box, click **Options** and in the next dialog box specify in **Means table: All terms in the model**. Minitab will display the estimated means for the significant effects in the output. Click **OK**, and in the main dialog box, choose **Graphs**. Under **Residual plots** choose **Four in one**. Minitab will display several residual plots to examine whether your model meets the assumptions of the analysis (Stat Tools 2.8 and 2.9). Click **OK** in the main dialog box.

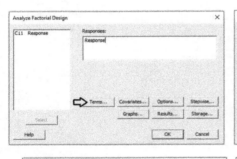

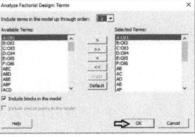

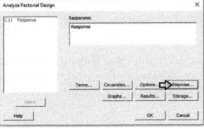

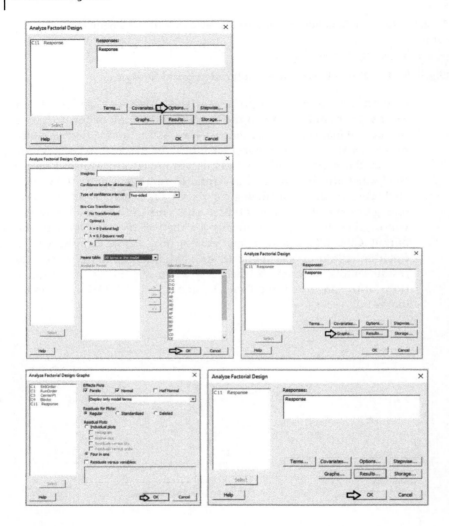

Complete the analysis by adding the factorial plots that show the relationships between the response and the significant terms in the model, thus displaying how the response mean changes as the factor levels or combinations of factor levels change.

To display factorial plots, go to:

Stat > DOE > Factorial > Factorial Plots

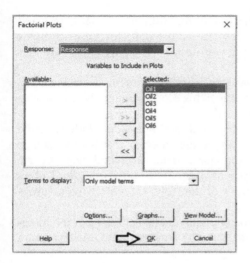

2.2.2.3.1 Interpret the Results of Step 3

Setting the significance level alpha to 0.05, the ANOVA table shows the significant terms in the model. Take into account that the stepwise procedure may add nonsignificant terms in order to create a hierarchical model. In a hierarchical model, all lower-order terms that comprise higher-order terms also appear in the model. You can see, for example, that the model includes the nonsignificant terms Oil4 and Oil6, because their interaction is present and also significant.

In addition to the results of the ANOVA, Minitab displays some other useful information in the Model Summary table.

Backward Elimination of Terms

α to remove = 0.05

Analysis of Variance

Source	DF	Adj SS	Adj MS	F-Value	P-Value
Model	6	16.7887	2.79812	4.80	0.000
Linear	4	9.2703	2.31758	3.98	0.006
Oil2	1	5.6870	5.68696	9.76	0.003
Oil4	1	0.4029	0.40288	0.69	0.409
Oil5	1	3.1466	3.14660	5.40	0.024
Oil6	1	0.0339	0.03386	0.06	0.810
2-Way Interactions	2	7.5184	3.75922	6.45	0.003
Oil4*Oil5	1	5.0718	5.07176	8.71	0.005
Oil4*Oil6	1	2.4467	2.44668	4.20	0.045
Error	57	33.1992	0.58244		
Total	63	49.9879			

Model Summary

S	R-sq	R-sq(adj)	R-sq(pred)
0.763179	33.59%	26.59%	16.27%

The quantity **R-squared** (R-sq, R^2) is interpreted as the percentage of the variability among liking scores, explained by the terms included in the ANOVA model.

The value of R^2 varies from 0% to 100%, with larger values being more desirable.

The adjusted R^2 (R-sq(adj)) is a variation of the ordinary R^2 that is adjusted for the number of terms in the model. Use adjusted R^2 for more complex experiments with several factors, when you want to compare several models with different numbers of terms.

The value of **S** is a measure of the variability of the errors that we make when we use the ANOVA model to estimate the liking scores. Generally, the smaller it is, the better the fit of the model to the data.

Before proceeding with the results reported in the session window, take a look at the residual plots. A **residual** represents an **error** that is the distance between an observed value of the response and its estimated value by the ANOVA model. The graphical analysis of residuals based on **residual plots** helps you to discover possible violations of the ANOVA underlying assumptions (Stat Tools 2.8 and 2.9).

A check of the normality assumption could be made by looking at the histogram of the residuals (on the lower left), but with small samples, the histogram often shows irregular shape.

The normal probability plot of the residuals may be more useful. Here we can see a tendency of the plot to bend upward slightly on the right, but the plot is not grossly non-normal in any case. In general, moderate departures from normality are of little concern in the ANOVA model with fixed effects.

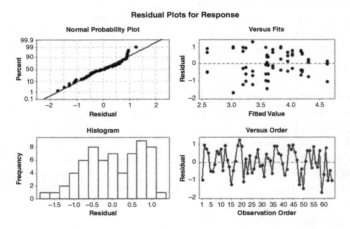

The other two graphs (Residuals vs. Fits and Residuals vs. Order) seem unstructured, thus supporting the validity of the ANOVA assumptions.

Returning to the Session window, we find a table of coded coefficients followed by a regression equation in uncoded units. To understand the meaning of these results and how we can interpret them, consider also the factorial plots and the Mean table.

In ANOVA, we have seen that many of the questions that the experimenter wishes to answer can be solved by several hypothesis tests. It is also helpful to present the results in terms of a **regression model**, i.e. an equation derived from the data that expresses the relationship between the response and the important design factors.

In the regression equation, the value 3.69 is the **intercept**. It's the sample response mean. We calculated it in the descriptive phase, considering all 65 observations. Then, in the regression equation we have a coefficient for each term in the model that indicates its impact on the response: a positive impact, if its sign is positive, or a negative impact, if its sign is negative.

In the regression equation, the coefficients are expressed in **uncoded units** – that is, in the original measurement units. These coefficients are also displayed in the Coded Coefficients table, but in **coded units**. What are the coded units? Minitab codes the low level of a factor to −1 and the high level of a factor to +1. The data expressed as −1 or +1 are in coded units.

Coded coefficients

Term	Effect	Coef	SE Coef	t-Value	p-Value	VIF
Constant		3.6901	0.0954	38.68	0.000	
Oil2	0.5962	0.2981	0.0954	3.12	0.003	1.00
Oil4	−0.1587	−0.0793	0.0954	−0.83	0.409	1.00
Oil5	−0.4435	−0.2217	0.0954	−2.32	0.024	1.00
Oil6	0.0460	0.0230	0.0954	0.24	0.810	1.00
Oil4*Oil5	−0.5630	−0.2815	0.0954	−2.95	0.005	1.00
Oil4*Oil6	0.3910	0.1955	0.0954	2.05	0.045	1.00

Regression Equation in Uncoded Units

$$\text{Response} = 3.6901 + 0.2981\ \text{Oil2} - 0.0793\ \text{Oil4} - 0.2217\ \text{Oil5} + 0.0230\ \text{Oil6} - 0.2815\ \text{Oil4*Oil5} + 0.1955\ \text{Oil4*Oil6}$$

When factors are categorical, the results in coded units or uncoded units are the same. When factors are quantitative, using coded units has a few benefits. One is that we can directly compare the size of the coefficients in the model, since they are on the same scale, thus determining which factor has the biggest impact on the response.

For example (for categorical factors):

Oil2	
Uncoded units	Coded units
NO	−1
YES	+1

Earlier, we learned how to examine the p-values in the ANOVA table to determine whether a factor is statistically related to the response. In the Coded Coefficients table, we find the same results in terms of tests on coefficients and we can evaluate the p-values to determine whether they are statistically significant. The null hypothesis is that a coefficient is equal to 0 (no effect on the response). Setting the significance level α at 0.05, if a p-value is less than the α level, reject the null hypothesis and conclude that the effect is statistically significant.

However, how can we interpret the regression coefficients? Looking at the Coded Coefficient table, see, for example, that when Oil2 is present (Oil2 = +1), the mean liking score increases by +0.298. This result is graphically presented in the factorial plots and with further details in the Means table. As Oil2 is not involved in a significant interaction, consider the Main Effects Plot.

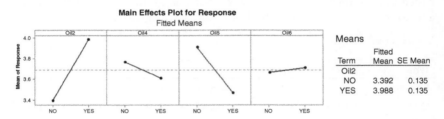

Means		
	Fitted	
Term	Mean	SE Mean
Oil2		
NO	3.392	0.135
YES	3.988	0.135

Extrapolate the plot related to Oil2 to analyze it in detail. The dashed horizontal line in the middle is the total mean liking score = 3.690. When Oil2 is present, the mean liking score increases by 0.298 (the magnitude of the coefficient) to 3.988 (this can be seen in the Means Table under Fitted Mean), while when absent, the mean liking score decreases by 0.298–3.392.

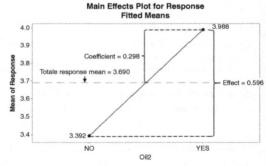

So the coefficient in coded units tells us the increase or the decrease with respect to the total mean.

Two times the coefficient is the effect of the factor (see its value in the Coded Coefficients table, under the column Effect).

For Oil5, we have the significance of its main effect and of the interaction between Oil5 and Oil4. The main effect of Oil4 is not statistically significant. The interaction between Oil4 and Oil6 is statistically significant. Look at the interaction plots and the Means table, and consider how the mean liking score varies as the levels of Oil4 and Oil5 change and as the levels of Oil4 and Oil6 change.

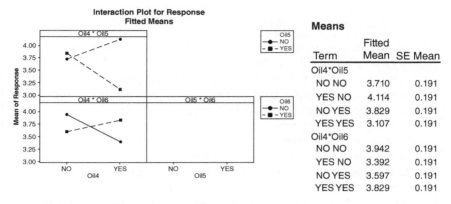

Interaction Plot for Response
Fitted Means

Means

Term	Fitted Mean	SE Mean
Oil4*Oil5		
NO NO	3.710	0.191
YES NO	4.114	0.191
NO YES	3.829	0.191
YES YES	3.107	0.191
Oil4*Oil6		
NO NO	3.942	0.191
YES NO	3.392	0.191
NO YES	3.597	0.191
YES YES	3.829	0.191

Considering the Oil4*Oil5 interaction, the best result (the highest mean liking score) is obtained when Oil4 is present and Oil5 is absent (fitted mean liking score = 4.114) and the worst result when both of them are present. For the Oil4*Oil6 interaction, the highest mean liking score is reached when both oils are absent (fitted mean liking score = 3.942), but the second best result is obtained when both oils are present. The interpretation of these results is not so clear. Remember that in a fractional design, we are not able to estimate all the main effects and interactions separately from each other: some of them will be confused (aliased) with others. In the fractional design we are analyzing, for example, factor B (Oil2) is aliased with the four-order interaction BCDEF; the ANOVA model estimates the effect of (B + ACDEF), but if we can reasonably assume that the interaction ACDEF is negligible, we can assign the total estimate to the main effect B. The three- and four-factor (higher-order) interactions are rarely considered because they represent complex events, which are sometimes difficult to interpret.

A three-factor interaction, for example, would mean that the effect of a factor changes as the setting of another factor changes, and that the effect of their two-factor interaction varies according to the setting of a third factor. As a consequence of this, out of the many factors that are investigated, we expect only a few to be statistically significant, and we can usually focus on the interactions containing factors whose main effects are themselves significant.

Returning to our example, based only on statistical considerations, it would seem reasonable to perform a new experiment to evaluate more deeply the effect of Oil2 and perhaps of Oil4 and Oil6. The likelihood of this interpretation should, however, be judged by the process experts. The experimenter might even think of adding some levels to the factors, considering, for example, three levels representing different concentrations of each oil. From a **two-level fractional design**, the researcher could move to a **general** (i.e. with more than two levels for at least one factor) **full or fractional design** to further investigate the relationship between oils and the liking scores.

Stat Tool 2.9 Residual Analysis

In the ANOVA procedure, for each group a *residual* represents an *error*, i.e. the distance between an observed value of the response and its value estimated by the ANOVA model, which is simply the mean response for that group. The *graphical analysis of residuals* based on *residual plots* helps us discover possible violations of the underlying ANOVA assumptions (Stat Tool 2.8).

Randomization and *random samples* (Stat Tools 2.3, 1.2) are prerequisites of the first assumption of *independent measurements*. We can plot the *residuals in the order that the data were collected* and look for any patterns. Structureless plots, showing no obvious patterns, reveal no serious problems of lack of independence. Look at the two *residuals* versus *order plots* in Figure 2.13. On the left, the points on the plot are randomly distributed and do not exhibit any patterns. The residuals appear to be *independent* of one another. The plot on the right, however, displays a recognizable *pattern*: from the 11th observation onwards, the points gradually increase over time, thus indicating that the residuals may not be independent.

A check of the *normality assumption* could be made through a *normal probability plot* of the residuals. If the underlying response distribution is normal, this plot will resemble a straight line. If the residuals approximately follow the *straight line on the plot*, you don't have a severe violation of the normality assumption. Below are two normal probability plots (Figure 2.14). The one on the left describes residuals that *likely follow a normal distribution*, where residuals follow approximately a straight line. The one on the right exhibits a pattern that indicates the residuals *do not follow a normal distribution*.

The *normal distribution* is bell-shaped and symmetric and is widely used in statistics to represent quantitative variables showing symmetry and unimodal distributions (Figure 2.15):

The third assumption states that the response *variance* across groups (treatment) is *constant* (*homoscedasticity*).

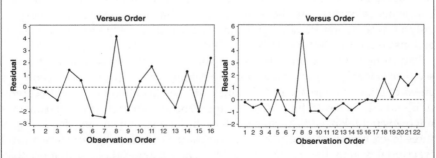

Figure 2.13 Residuals versus order plots.

Stat Tool 2.9 (Continued)

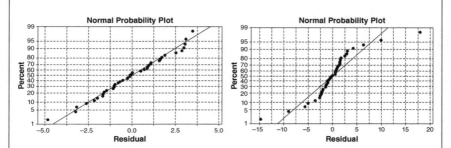

Figure 2.14 Normal probability plots.

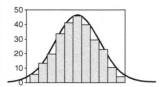

Figure 2.15 Normal distribution.

To evaluate this assumption, we can *plot the residuals* versus *the fitted values*. Notice that the residual values are on the vertical axis and the fitted values are on the horizontal axis.

Remember that the fitted values are obtained using the ANOVA model and are simply the mean response values for each group. Of the two plots in Figure 2.16, the one on the left shows *constant variance across groups*: the scatter of the points is similar across all groups and shows no patterns. In the plot on the right, the residuals increase systematically with the fitted values, which is sometimes called the funnel effect. Here, the *variance is not constant across groups*.

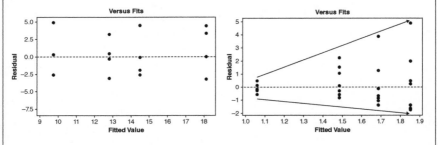

Figure 2.16 Residuals versus fitted values plots.

3

Product Development and Optimization

3.1 Introduction

Developing or improving products, streamlining a production process or comparing the performance of alternative formulations, very often represent core issues for researchers and managers (Box, G. E. P. and Woodall, W. H., 2012; Hoerl, R., Snee, R., 2010; Jensen, W. et al., 2012).

In this context, many questions can be viewed as problems of comparison of synthetic measures, such as mean values or proportions, with expected target values, or of comparison among different configurations of the investigated product. Furthermore, statistical modeling and optimization techniques find a vast number of applications when the objective is to study in depth how one or more desirable product characteristics are related to a set of process variables and how to set these variables in order to optimize product performance.

In this chapter, four different case studies are presented to cover several concrete situations that experimenters can encounter in the development and optimization phases.

The first example refers to the development of a new sore throat medication, where the research team aims to check the following issues:

- The mean content of an anesthetic ingredient is not different to the target value of 2.4 mg.
- The percentage of lozenges with total weight greater than 2.85 mg is less than 40%.

One-sample inferential technique allows us to compare sample means or proportions to target values.

In the second case study, two different formulations of condoms have been manufactured and the researchers would like to evaluate the following features:

End-to-End Data Analytics for Product Development: A Practical Guide for Fast Consumer Goods Companies, Chemical Industry and Processing Tools Manufacturers, First Edition.
Rosa Arboretti, Mattia De Dominicis, Chris Jones, and Luigi Salmaso.
© 2020 John Wiley & Sons Ltd. Published 2020 by John Wiley & Sons Ltd.
Companion website: www.wiley.com/go/salmaso/data-analytics-for-pd

- Is there any significant difference in the *variability* of thickness between the two formulations?
- Is there any significant difference in the *mean* of thickness between the two formulations?
- Is there any significant difference in the *proportion* of condoms with thickness less than or equal to 0.045 mm between the two formulations?

For these kinds of problems, two-sample inferential techniques can help the investigators to identify the better formulation.

The third example, the Fragrance Project, aims to introduce an innovative shaving oil. It is of interest to assess two fragrances to investigate whether there are any consumer perceived differences in their in-use characteristics and if so, which is the more suitable. A total of 30 female respondents are given the two fragrances and asked to answer how appropriate the fragrance is for a shaving product by assigning an integer score from 0 (very unsuitable) to 10 (very suitable). The research team needs to assess whether there are any significant differences in appropriateness between the two fragrances. Since the respondents evaluate both fragrances, a specific hypothesis test for dependent samples (paired data) must be used.

The final example refers to a Stain Removal Project that aims to complete the development of an innovative dishwashing product. During a previous screening phase, investigators selected four components affecting cleaning performance by using a two-level factorial design. After identifying the key input factors, the experimental work proceeds with a general factorial design increasing the number of levels of the four components in order to identify the formulation that maximizes stain release on several stains. For logistic reasons, investigators need to arrange the new experiment in such a way that only 40 different combinations of factor levels will be tested. In this context, the planning and analysis of general and optimal factorial designs (Ronchi, F., et al., 2017) allows us to identify which formula could maximize cleaning performance, taking into account the constraint in the number of runs.

In short, the chapter deals with the following:

Topics	Stat Tools
Comparison of a mean to a target value: One-sample t-test	3.1
Comparison of a proportion to a target value: One proportion test	3.2
Comparison of two variances: Two variances test	3.3, 3.4
Comparison of two means with independent samples: Two-sample t-test	3.3, 3.5
Comparison of two means with paired data: Paired t-test	3.7
Comparison of two proportions: Two proportions test	3.6
Response optimization in factorial designs	3.8

Learning Objectives and Outcomes

Upon completion of this chapter, you should:

Know how to solve one-sample and two-sample inferential problems for the development of products/services.

Be able to conduct a one-sample t-test for the population mean to compare it to a target value.

Understand how the one proportion test can be used to assess whether a population proportion differs from a target value.

Be able to conduct a two variances test to compare two population variances.

Be able to conduct a two-sample t-test to compare two population means based on two independent samples or with paired data.

Be able to conduct a two proportions test to compare two population proportions.

Know how to reduce the number of runs of a full factorial design by selecting an optimal design.

Understand how to evaluate the relationship between a set of factors and one or more responses in factorial designs.

Be able to detect factor settings that optimize one or more responses.

3.2   Case Study for Single Sample Experiments: Throat Care Project

In the development of a new throat medication, the research team aims to check the following issues:

- The mean content of an anesthetic ingredient does not differ from the target value of 2.4 mg.
- The percentage of lozenges with total weight greater than 2.85 mg is less than 40%.

Under controlled conditions, 30 trials were performed by randomly selecting a sample of 30 lozenges (experimental units, Stat Tool 1.2) and measuring the content of the ingredient and the total weight of each lozenge (statistical variables, Stat Tool 1.1).

In the Throat Care project, investigators must deal with *one-sample inferential problems* as it is of interest to compare a population parameter (e.g. a mean value or a proportion) with a specified value (a target, standard, or historical value).

To solve these problems, we can apply an appropriate *one-sample hypothesis test*:

- Considering the variable "content of the anesthetic ingredient," use the *one-sample t-test* to determine whether the *population mean content* is not different to the required value of 2.4 mg.
- Considering the variable "total weight," use the *one proportion test* to determine whether the *population proportion of lozenges* with total weight greater than 2.85 mg is less than 0.40.

The variable settings are the following:

- Variable "Ingredient" is a *continuous quantitative variable* expressed in mg.
- Variable "Weight" is a *continuous quantitative variable* expressed in mg.

Column	Variable	Type of data	Label
C1	Lozenges		Lozenges' code
C2	Ingredient	Numeric data	Content of the anesthetic agent in mg
C3	Weight	Numeric data	Total weight of each lozenge in mg

File: Throat_Care_Project.xlsx

3.2.1 Comparing the Mean to a Specified Value

To determine whether the *population mean content* does not differ from the required value of 2.4 mg, let's proceed as follows:

Step 1 – Perform a descriptive analysis (Stat Tool 1.3) of the variable "Ingredient."
Step 2 – Assess the null and the alternative hypotheses (Stat Tool 1.15) and apply the one-sample t-test (Stat Tools 1.16, 3.1).

3.2.1.1 Step 1 – Perform a Descriptive Analysis of the Variable "Ingredient"

For the quantitative variable "Ingredient," use the dotplot (recommended for small samples with less than 50 observations) or the histogram (for moderate or large datasets with sample size greater than 20) to show its distribution. Add the boxplot and calculate means and measures of variability to complete its descriptive analysis.

 To display the histogram, go to: **Graph > Histogram**
Check the graphical option **Simple** and in the next dialog box, select "Ingredient" in **Graph variables**. Then, select the option **Scale** and in the next screen choose **Y-scale type**, to display percentages in the histogram. Then click **OK** in the main dialog box.

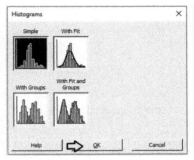
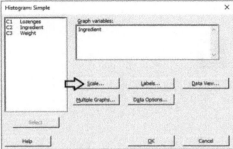

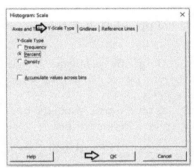
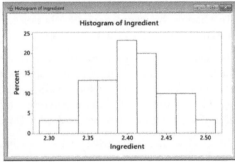

To change the width and number of bar intervals, double-click on any bar and select the **Binning** tab. A dialog box will open with several options to define. Specify 5 in **Number of intervals** and click **OK**.

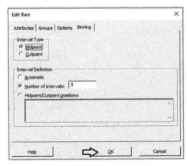
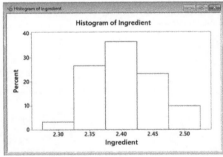

Use the tips in Chapter 2, Section 2.2.2.1 to change the appearance of the histogram.

To display the boxplot, go to: **Graph** > **Boxplot**

Check the graphical option **Simple** and in the next screen, select "Ingredient" in **Graph variables**. Then click **OK** in the main dialog box.

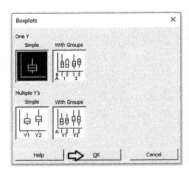 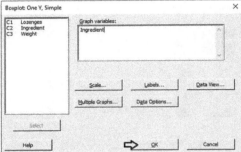

To change the appearance of a boxplot, use the tips previously referred to histograms, as well as the following:

- To display the boxplot horizontally: double-click on any scale value on the horizontal axis. A dialog box will open with several options to define. Select from these options: Transpose value and category scales.
- To add the mean value to the boxplot: right-click on the area inside the border of the boxplot, then select **Add > Data display > Mean symbol**.

To display the descriptive measures (means, measures of variability, etc.), go to:

Stat > Basic Statistics > Display Descriptive Statistics

Select "Ingredient" in **Variables**; then click on **Statistics** to open a dialog box displaying a range of possible statistics to choose from. In addition to the default options, select **Interquartile range** and **Range**.

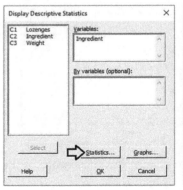 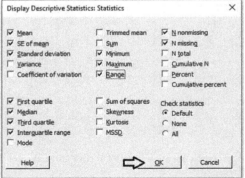

3.2.1.1.1 Interpret the Results of Step 1 Let us describe the shape, central and non-central tendency and variability of the content of the anesthetic ingredient.

Shape: the distribution is fairly symmetric: middle values are more frequent than low and high values (Stat Tools 1.4–1.5, 1.11).

Central tendency: the mean is equal to 2.4058 mg. Mean and median are close in values (Stat Tools 1.6, 1.11).

Non-central tendency: 25% of the values are less than 2.3706 mg (first quartile Q_1) and 25% are greater than 2.4402 mg (third quartile Q_3), (Stat Tools 1.7, 1.11).

Variability: observing the length of the boxplot (range) and the width of the box (interquartile range, IQR), the distribution shows moderate variability. The lozenges' content varies from 2.2953 mg (minimum) to 2.4976 mg (maximum). 50% of middle values vary from 2.3706 to 2.4402 mg (IQR = maximum observed difference between middle values equal to 0.0696 mg) (Stat Tools 1.8, 1.11).

The average distance of the values from the mean (standard deviation) is 0.0486 mg: the lozenges' content varies on average from (2.4058–0.0486) = 2.3572 mg to (2.4058 + 0.0486) = 2.4544 mg (Stat Tools 1.9).

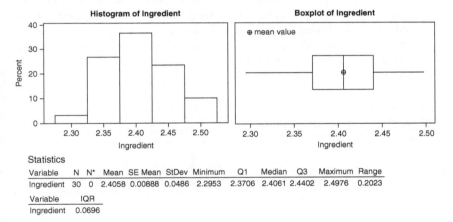

Statistics

Variable	N	N*	Mean	SE Mean	StDev	Minimum	Q1	Median	Q3	Maximum	Range
Ingredient	30	0	2.4058	0.00888	0.0486	2.2953	2.3706	2.4061	2.4402	2.4976	0.2023

Variable	IQR
Ingredient	0.0696

From a descriptive point of view, the mean content of the anesthetic agent seems not to be different from the required value of 2.4.

Is the difference between the sample mean of 2.4058 and the target value 2.4 statistically significant or NOT? Do the sample results lead to rejection of the null hypothesis of equality of the population mean content of the anesthetic ingredient to the required value? Apply the one-sample t-test to answer these questions.

3.2.1.2 Step 2 – Assess the Null and the Alternative Hypotheses and Apply the One-Sample t-Test

The appropriate procedure to determine whether the *population mean content* does not differ from the required value of 2.4 mg is the one-sample t-test (Stat Tool 3.1). What are the *null* and *alternative hypotheses*?

Null hypothesis H_0: Mean content of the anesthetic ingredient = 2.4
Alternative hypothesis H_1: Mean content of the anesthetic ingredient ≠ 2.4

Look at the alternative hypothesis: the mean content can be greater or less than 2.4. This is a bidirectional alternative hypothesis. In this case, the test is called a *two-tailed* (or *two-sided*) *test*. Next, we'll look at the test results and decide whether to reject or fail to reject the null hypothesis.

 To apply the one-sample t-test, go to:

Stat > Basic Statistics > 1 – Sample t

In the dialog box, leave the option **One or more samples, each in a column**, and select the variable "Ingredient." Check the option **Perform hypothesis test** and specify the target value "2.4" in **Hypothesized mean**. Click **Options** if you want to modify the significance level of the test or specify a directional alternative. Click on **Graphs** to display the boxplot with the null hypothesis and the confidence interval for the mean.

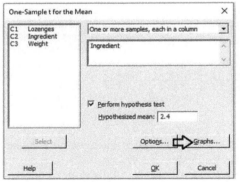
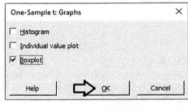

3.2.1.2.1 Interpret the Results of Step 2   The sample mean 2.4058 mg is a point estimate of the population mean content of the anesthetic agent. The confidence interval provides a range of likely values for the population mean. We can be 95% confident that on average the true mean content is between 2.3876 and 2.4239. Notice that the CI includes the value 2.4. Because the p-value (0.522) is greater than the significance level α = 0.05, we fail to reject the null hypothesis.

There is not enough evidence that the mean content is different from the target value 2.4 mg. The difference between the sample mean 2.4058 mg and the required value 2.4 mg is *not statistically significant*.

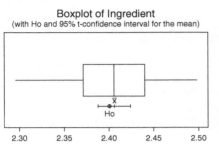

Boxplot of Ingredient
(with Ho and 95% t-confidence interval for the mean)

One-Sample T: Ingredient
Descriptive Statistics

N	Mean	StDev	SE Mean	95% CI for μ
30	2.40576	0.04864	0.00888	(2.38760; 2.42392)

μ: mean of Ingredient

Test

Null hypothesis	H_0: μ = 2.4
Alternative hypothesis	H_1: μ ≠ 2.4

T-Value	P-Value
0.65	0.522

Stat Tool 3.1 One-Sample t-Test

A common task in statistical studies is the *comparison of the mean* to a hypothesized value of interest.

Considering a *quantitative variable Y*, use the *one-sample t-test* to determine whether the population mean μ is equal to a hypothesized value μ_0.

The one-sample t-test requires a *random sample* selected from a population (Stat Tool 1.2). We use the *sample mean* to draw conclusions about the *population mean*.

Population
Population mean of Y = μ ⟶ μ_0

Random sample
Sample mean of Y = $\bar{x}$ ⟶ μ_0

➤ *Example 3.1*. It is of interest to evaluate if the *time for water repellency* of a new formulation of a stain removal product is greater than 100 seconds (hypothesized value).

The one-sample t-test considers the sample mean of the times for water repellency measured for a random sample of fabric specimens treated with the new formulation.

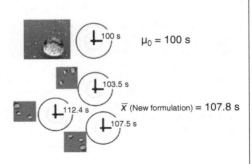

μ_0 = 100 s

$\bar{x}$ (New formulation) = 107.8 s

Stat Tool 3.1 (Continued)

It evaluates the difference between the sample mean and the hypothesized value:

$\bar{x}$ (New formulation) $-$ 100 s (μ_0)

and allows us to draw conclusions about the difference between population mean and the hypothesized value:

μ (New formulation) $-$ 100 s (μ_0)

by determining whether this mean is:

μ (New formulation) $\neq$ 100 s (μ_0)

- Statistically significant: there is a difference between the population mean and the hypothesized value.

μ (New formulation) $=$ 100 s (μ_0)

- NOT statistically significant: there is no difference between the population mean and the hypothesized value.

The one-sample t-test results in Minitab include both a confidence interval and a p-value:

- Use the confidence interval (usually a 95% or a 99% CI) to obtain a range of reasonable values for the population mean: μ. The central point of the confidence interval is the sample mean: $\bar{x}$ (point estimate of μ).
- Use the p-value to determine whether the population mean differs from the hypothesized value μ_0.

For a one-sample t-test:

Null Hypothesis:	**Alternative Hypothesis:**
Population mean is EQUAL to the hypothesized value.	Population mean is NOT EQUAL to the hypothesized value.
$H_0: \mu = \mu_0$	$H_1: \mu \neq \mu_0$ or $H_1: \mu < \mu_0$ or $H_1: \mu > \mu_0$

Setting the significance level α usually at 0.05 or 0.01:

- If the p-value is less than α, reject the null hypothesis and conclude that the difference between the mean and the hypothesized value is statistically significant. | p-value < α | Reject the null hypothesis that population mean equals μ_0.
- If the p-value is greater than or equal to α, fail to reject the null hypothesis and conclude that the difference between the mean and the hypothesized value is NOT statistically significant. | p-value $\geq$ α | Fail to reject the null hypothesis that population mean equals μ_0.

3.2.2 Comparing a Proportion to a Specified Value

To determine whether the *population proportion of lozenges* with total weight greater than 2.85 mg is less than 0.40, let's proceed in the following way:

Step 1 – Calculate the sample proportion of lozenges with total weight greater than 2.85 mg.
Step 2 – Assess the *null* and the *alternative hypotheses* (Stat Tool 1.15) and apply the *one proportion test* (Stat Tools 1.16, 3.2).

3.2.2.1   Step 1 – Calculate the Sample Proportion of Lozenges with Total Weight Greater than 2.85 mg

Let us create a new categorical variable from the total weight that incorporates the two categories: "total weight less than or equal to 2.85," "total weight greater than 2.85."

To create a new categorical variable, go to: **Data > Recode > To Text**

Under **Recode values in the following columns**, select the variable "Weight." Then, under **Method**, choose the option **Recode ranges of values**. Define the lower and upper endpoint of each category and in **Endpoints to include**, specify **Lower endpoint only**. Then click **OK** in the dialog box.

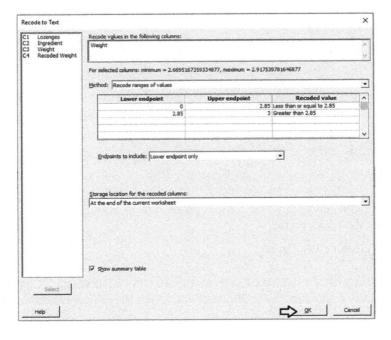

Note that the new variable "Recoded Weight" will appear at the end of the current worksheet. This variable is a categorical variable incorporating two categories: "Less than or equal to 2.85," "Greater than 2.85."

	C1	C2	C3	C4-T	C5	C6
	Lozenges	Ingredient	Weight	Recoded Weight		
1	1	2.487	2.846	Less than or equal to 2.85		
2	2	2.389	2.832	Less than or equal to 2.85		
3	3	2.346	2.759	Less than or equal to 2.85		
4	4	2.383	2.813	Less than or equal to 2.85		
5	5	2.295	2.867	Greater than 2.85		
6	6	2.395	2.854	Greater than 2.85		

To calculate the sample proportion of lozenges with total weight greater than 2.85, go to:

Stat > Tables > Tally Individual variables

In **Variables**, select the variable "Recoded Weight." Then, in **Display**, check the options **Count** and **Percents**. Click **OK** in the dialog box.

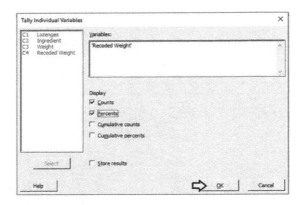

To graphically display the sample proportion of lozenges with total weight greater than 2.85, go to:

Graph > Pie Chart

In **Categorical variables**, select the variable "Recoded Weight." Click **Labels** and in the dialog box select **Slice Labels**. Check the option **Percent** and click **OK** in each dialog box.

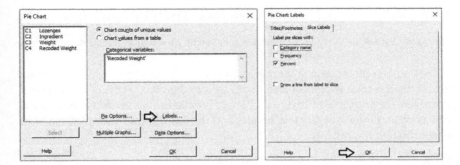

3.2.2.1.1 Interpret the Results of Step 1 The table stored in the Session window and the pie chart show the frequency distribution of the variable "Recoded weight." In 7 out of 30 lozenges (sample percentage = 23.3%; sample proportion = 0.233), the total weight is greater than 2.85 mg.

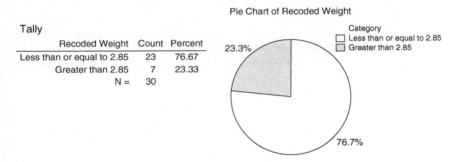

From a *descriptive point of view*, the new formulated lozenges seem not to have exceeded the critical value of 40%, as the sample percentage is equal to 23.3%.

Is the difference between the sample proportion of 0.233 and the critical proportion of 0.40 statistically significant or NOT? Do the sample results lead to rejection of the null hypothesis of equality of the population proportion to the critical value? Apply the one proportion test to answer these questions.

3.2.2.2 Step 2 – Assess the Null and the Alternative Hypotheses and Apply the One Proportion Test

The appropriate procedure to determine whether the proportion of lozenges with total weight greater than 2.85 mg is less than 0.40 is the One Proportion Test (Stat Tool 3.2). What are the *null* and *alternative hypotheses*?

Null hypothesis	H_0: Proportion of lozenges with total weight greater than 2.85 mg = 0.40.
Alternative hypothesis	H_1: Proportion of lozenges with total weight greater than 2.85 mg < 0.40.

Look at the alternative hypothesis: the researchers expect the proportion exceeding the total weight of 2.85 mg to be less than 0.40. This is a directional alternative hypothesis. In this case the test is called a *one-tailed* (or *one-sided*) *test*. Next, we'll look at the test results and decide whether to reject or fail to reject the null hypothesis.

To apply the one proportion test, go to:

Stat > Basic Statistics > 1 Proportion

In the drop-down menu at the top, choose **Summarized data**. In **Number of events**, enter 7 (the number of lozenges exceeding 2.85 mg as total weight). In **Number of trials**, enter 30 (the sample size). Check **Perform hypothesis test**. In **Hypothesized proportion**, enter 0.40 (the critical value). Click **Options**. Under **Alternative hypothesis**, choose **Proportion < hypothesized proportion**, and under **Method** select **Normal approximation**. If the number of events and non-events is less than 5 observations, under **Method** choose **Exact**. Click **OK** in each dialog box.

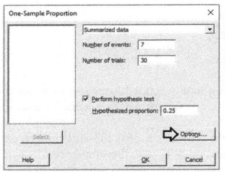
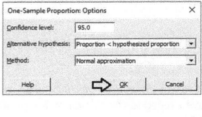

3.2.2.2.1 Interpret the Results of Step 2 The sample proportion 0.233 is a point estimate of the population proportion of lozenges with total weight greater than 2.85 mg. In our case, we required a one-sided test; therefore, only the lower bound of the confidence interval is shown. We can be 95% confident that the true proportion is less than 0.36. Notice that the lower bound is less than the critical value 0.40. Because the p-value (0.031) is less than the significance level $\alpha = 0.05$, we reject the null hypothesis. There is enough evidence that the proportion is less than 0.40. The difference between the sample proportion 0.233 and the critical value 0.40 is *statistically significant*.

Test and CI for One Proportion
Method

p: event proportion
Normal approximation method is used for this analysis.

Descriptive Statistics

N	Event	Sample p	95% Upper Bound for p
30	7	0.233333	0.360349

Test

Null hypothesis	$H_0: p = 0.4$
Alternative hypothesis	$H_1: p < 0.4$

z-Value	p-Value
−1.86	0.031

Stat Tool 3.2 One Proportion Test

Another common task in statistical studies is the comparison between the *proportion* of statistical units with a characteristic of interest, and a hypothesized value.

Use the *one proportion test* to determine whether the population proportion is equal to the hypothesized value. The one proportion test requires a *random sample* selected from a population (Stat Tool 1.2). We use the *sample proportion* to draw conclusions about the *population proportion*.

Population
Population proportion = π ⟶ π_0

Random sample
Sample proportion = p ⟶ π_0

➤ *Example 3.2.* It is of interest to evaluate whether the proportion of cotton fabrics that, after the application of a new formulation of a stain removal product, have had the stain completely repelled, is equal to 20% (hypothesized value).

$\pi_0 = 0.20$

The one proportion test considers the sample proportion (relative frequency) with a completely repelled stain, in a random sample of fabric specimens treated with the new formulation.

p (New formulation) = 2/9 = 0.22

Stat Tool 3.2 (Continued)

It evaluates the difference between the sample proportion and the hypothesized value:

and allows us to draw conclusions about the difference between population proportion and the hypothesized value:

by determining whether this difference is:

$p_{\text{(New formulation)}} - 0.20\ (\pi_0)$

$\Downarrow$

$\pi_{\text{(New formulation)}} - 0.20\ (\pi_0)$

$\pi_{\text{(New formulation)}} \neq 0.20\ (\pi_0)$

$\pi_{\text{(New formulation)}} = 0.20\ (\pi_0)$

- Statistically significant: There is a difference between the population proportion and the hypothesized value.
- NOT statistically significant: There is no difference between the population proportion and the hypothesized value.

The one proportion test results in Minitab include both a confidence interval and a p-value:

- Use the confidence interval (usually a 95% or a 99% CI) to obtain a range of reasonable values for the population proportion: π. The central point of the confidence interval is the sample proportion: p (point estimate of π).
- Use the p-value to determine whether the population proportion differs from the hypothesized value π_0.

For a one proportion test:

Null Hypothesis:	**Alternative Hypothesis:**
Population proportion is EQUAL to the hypothesized value.	Population proportion is NOT EQUAL to the hypothesized value.
$H_0: \pi = \pi_0$	$H_1: \pi \neq \pi_0$ or $H_1: \pi < \pi_0$ or $H_1: \pi > \pi_0$

Setting the significance level α usually at 0.05 or 0.01:

• If the p-value is less than α, reject the null hypothesis and conclude that the difference between the proportion and the hypothesized value is statistically significant.	p-value $< \alpha$	Reject the null hypothesis that population proportion equals $\pi0$.
• If the p-value is greater than or equal to α, fail to reject the null hypothesis and conclude that the difference between the proportion and the hypothesized value is NOT statistically significant.	p-value $\geq \alpha$	Fail to reject the null hypothesis that population proportion equals $\pi0$.

3.3   Case Study for Two-Sample Experiments: Condom Project

Two different formulations (A, B) of condoms have been manufactured. The team would like to know if there is any significant difference between the thickness of the condoms and have provided data from across the product. Under controlled conditions, 48 trials were performed by randomly selecting a sample of 23 condoms for formulation A and 25 condoms for formulation B and measuring the thickness at 30 + 5 mm from open end.

They need to assess the following issues:

- Is there any significant difference in the *variability* of thickness between the two formulations?
- Is there any significant difference in the *mean* of thickness between the two formulations?
- Is there any significant difference in the *proportion* of condoms with thickness less than or equal to 0.045 mm, between the two formulations?

In the Condom Project, investigators must deal with *two-sample inferential problems* (Stat Tool 3.3) as the comparison of a population parameter (e.g. a mean value, a measure of variability or a proportion) between two groups of statistical units is of interest.

To solve these problems, we can apply an appropriate *two-sample hypothesis test.*

In relation to the variable "thickness at 30 + 5 mm from open end:"

- Use the *two variances test* (Stat Tool 3.4) to determine whether the *population variability* differs between the two formulations.
- Use the *two-sample t-test* (Stat Tool 3.5) to determine whether the *population mean* differs between the two formulations.
- Use the *two proportions test* (Stat Tool 3.6) to determine whether the *population proportion of condoms* with a thickness at 30 + 5 mm from open end of less than or equal to 0.045 mm, differs between the two formulations.

The variable settings are the following:

- The variable "Open_end" is a *continuous quantitative variable,* expressed in mm.
- The variable "Formulation" is a *categorical variable,* assuming two categories: "A" and "B."

Column	Variable	Type of data	Label
C1	Condom		Condoms' code
C2	Formulation	Categorical data	Formulations A and B
C3	Open_end	Numeric data	Thickness at 30 + 5 mm from open end, in mm

File: Condom_Project.xlsx

3.3.1 Comparing Variability Between Two Groups

To determine if there is any significant difference in the *variability* of thickness at 30 + 5 mm from open end between the two formulations, let's proceed in the following way:

Step 1 – Perform a descriptive analysis (Stat Tool 1.3) of the variable "Open_end" stratifying by formulations.

Step 2 – Assess the null and the alternative hypotheses (Stat Tools 1.15) and apply the Two Variances Test (Stat Tools 1.16, 3.4).

3.3.1.1 Step 1 – Perform a Descriptive Analysis of the Variable "Open_end" Stratifying by Formulations

For the quantitative variable "Open_end," use the dotplot (recommended for small samples with less than 50 observations) or the histogram (for moderate or large datasets with sample size greater than 20), stratifying by formulations. Finally, add the boxplot (for moderate or large datasets) and calculate means and measures of variability to complete the descriptive analysis.

 To display the dot plots by formulations, go to: **Graph** > **Dotplot** Under **One Y**, select the graphical option **With Groups** and in the next dialog box, select "Open_end" in **Graph variables** and "Formulation" in **Categorical variables for grouping**. Then click **OK** in the dialog box.

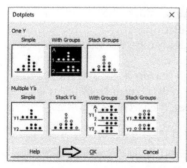 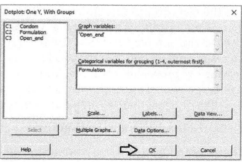

To display the boxplots by formulations, go to: **Graph** > **Boxplot**

Under **One Y**, select the graphical option **With Groups** and in the next dialog box, select "Open_end" in **Graph variables** and "Formulation" in **Categorical variables for grouping**. Then click **OK** in the dialog box. To change the appearance of a boxplot, use the tips in section 3.2.1.1.

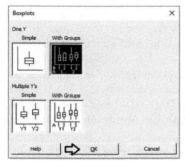 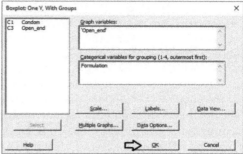

To display the descriptive measures (means, measures of variability, etc.) by formulations, go to:

Stat > Basic Statistics > Display Descriptive Statistics

Select "Open_end" in **Variables**, and "Formulation" in **By variables**, then click on **Statistics** to open a dialog box displaying a range of possible statistics to choose from. In addition to the default options, select **Variance, Coefficient of variation, Interquartile range** and **Range**. Click **OK** in each dialog box.

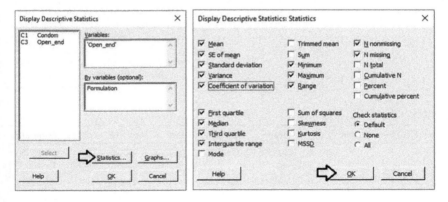

3.3.1.1.1 Interpret the Results of Step 1 Let's describe the shape, central, and non-central tendency and variability of the thickness at 30 + 5 mm from open end for the two formulations.

Shape: The distribution is asymmetric to the left for formulation A while formulation B tends to be more symmetric (Stat Tools 1.4, 1.5, 1.11).

Central tendency: The mean is equal to 0.046 mm for group A and 0.048 for group B. Mean and median are close in values for formulation B (Stat Tools 1.6, 1.11).

Non-central tendency: 25% of condoms of type A have thickness less than 0.044 mm (first quartile Q_1) and 25% greater than 0.048 mm (third quartile Q_3), (Stat Tools 1.7, 1.11). For formulation B, the first and third quartiles are respectively equal to 0.046 mm and 0.050 mm.

Variability: Observing the length of the boxplot (range) and the width of the box (Interquartile range, IQR), formulation B shows slightly higher variability than formulation A. Condom thickness varies from 0.041 mm (minimum) to 0.049 mm (maximum) for group A and from 0.041 mm (minimum) to 0.055 mm (maximum) for group B (Stat Tools 1.8, 1.11).

For formulation A, the average distance of the values from the mean (standard deviation) is 0.002 mm: condom thickness varies on average from (0.046–0.002) = 0.044 mm to (0.046 + 0.002) = 0.048 mm (Stat Tools 1.9). For formulation B, the standard deviation is 0.003 mm: condom thickness varies on average from 0.045 to 0.051 mm (mean ± standard deviation). Remember that the standard deviation is the square root of the variance.

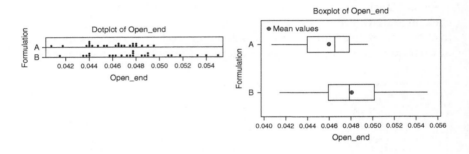

Statistics

Variable	Formulation	N	N*	Mean	SE Mean	StDev	Variance	CoefVar	Minimum
Open _end	A	23	0	0.045961	0.000479	0.002298	0.000005	5.00	0.040666
	B	25	0	0.048029	0.000680	0.003398	0.000012	7.07	0.041419

Variable	Formulation	Q1	Median	Q3	Maximum	Range	IQR
Open _end	A	0.043959	0.046498	0.047853	0.049540	0.008874	0.003894
	B	0.045901	0.047852	0.050127	0.055029	0.013610	0.004225

From a *descriptive point of view*, formulation B seems to have slightly higher variability than A. Look at the coefficient of variation (Stat Tool 1.10), which is greater for condoms of type B. The two standard deviations (or variances) seem to be different. Is this difference statistically significant or NOT? Do the sample results lead to rejection of the null hypothesis of equality of the population variances for the two formulations? Apply the two variances test to answer these questions.

3.3.1.2 Step 2 – Assess the Null and the Alternative Hypotheses and Apply the Two Variances Test

The appropriate procedure to determine whether the variability of thickness measured at $30 + 5$ mm from open end differs between the two formulations is the two variances test (Stat Tool 3.4). What are the *null* and *alternative hypotheses*?

Null hypothesis

H_0: For the two formulations, the variances of thickness at $30 + 5$ mm from open end are equal.

Alternative hypothesis

H_1: For the two formulations, the variances of thickness at $30 + 5$ mm from open end are not equal.

Look at the alternative hypothesis: the group A variance can be greater or less than the group B variance. This is a bidirectional alternative hypothesis. In this case the test is called a *two-tailed* (or *two-sided*) *test*. Next we'll look at the test results and decide whether to reject or fail to reject the null hypothesis.

 To apply the two variances test, go to:

Stat > Basic Statistics > 2 variances

> In the drop-down menu at the top, choose **Both samples are in one column**. In **Samples**, select the variable "Open_end" and in **Sample IDs**, select the variable "Formulation," then click **Options**. Note that the **2 variances** procedure performs hypothesis tests and computes confidence intervals for the *ratios* between two populations' variances or standard deviations. This is why the alternative hypothesis is related to the ratio of the two populations' variances or standard deviations, where a ratio of 1 suggests equality between populations. Under **Ratio**, leave the selection **(sample 1 standard deviation) / (sample 2 standard deviation)**. Under **Alternative hypothesis**, choose **Ratio ≠ hypothesized ratio**, and check the option **Use test and confidence intervals based on normal distribution**. If your samples have less than 20 observations each, or the distribution for one or both populations is known to be far from the normal distribution, uncheck this option.
> Click **OK** in each dialog box.

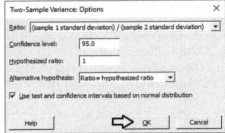

3.3.1.2.1 Interpret the Results of Step 2

The sample ratio of standard deviations 0.676 is a point estimate of the population ratio. The confidence interval provides a range of likely values for the population ratio. We can be 95% confident that on average the true ratio is between 0.446 and 1.033. Notice that the CI includes the value 1. Because the p-value (0.069) of the test is greater than the significance level $\alpha = 0.05$, we fail to reject the null hypothesis. There is insufficient evidence that the variability is different between the two formulations. The difference between the two standard deviations (or variances) is *not statistically significant.*

Test and CI for Two Variances: Open_end versus Formulation

Method

σ_1: standard deviation of Open_end when Formulation = A
σ_2: standard deviation of Open_end when Formulation = B
Ratio: σ_1/σ_2
F method was used. This method is accurate for normal data only.

Descriptive Statistics

Formulation	N	StDev	Variance	95% CI for σ
A	23	0.002	0.000	(0.002; 0.003)
B	25	0.003	0.000	(0.003; 0.005)

Ratio of Standard Deviations

Estimated ratio	95% CI for ratio using F
0.676203	(0.446; 1.033)

Test

Null hypothesis	H_0: $\sigma_1 / \sigma_2 = 1$
Alternative hypothesis	H_1: $\sigma_1 / \sigma_2 \neq 1$
Significance level	$\alpha = 0.05$

Method	Test Statistic	DF1	DF2	p-Value
F	0.46	22	24	0.069

Stat Tool 3.3 Two-Sample Inferential Problems

Earlier we studied experiments in which the mean or a proportion of a single population was compared against standard or target values. Another frequent research interest is the comparison of mean values, measures of variability (variances or standard deviations) or proportions between *two groups*. E.g.

It is of interest to compare the mean blood pressure between a test drug and the placebo.

It is of interest to compare the variability (e.g. the variances) of blood pressure between a test drug and the placebo.

It is of interest to compare the proportions of patients with blood pressure (systolic) less than 120 mm Hg, between a test drug and the placebo.

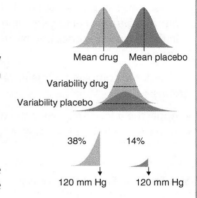

Let's consider the comparison of the *mean values* of a quantitative variable between two groups.

When you conduct a hypothesis test on means using *two random samples* (Stat Tool 1.2) from the two populations to compare, you must choose the type of test based on whether the samples are *dependent* or *independent*:

- If sampling units are the same in the two samples, and they are observed in two different conditions, then the samples are *dependent*.
- If sampling units are different between two samples and they are observed in two different conditions, then the samples are *independent*.

First sample Second sample

Dependent data (also known as *paired data*) usually arise when we observe or measure categorical or numerical variables *on the same individual or statistical unit at different points in time/ space*.

Before After

Without drug With drug

Let's consider the following example on collecting dependent and independent samples.

First sample Second sample

> *Example 3.3.* You need to test the effectiveness of a new drug in reducing blood pressure. You could collect data in two ways:

Placebo Active drug

- Sample the blood pressures of the *same people before and after* they receive a dose. The two samples are *dependent* because they are taken from the same

Stat Tool 3.3 (Continued)

people. The people with the highest blood pressure in the first sample will likely have the highest blood pressure in the second sample.

- Give one group of people an active drug and give a different group of people an inactive placebo, then compare the blood pressures between the groups. These two samples would likely be *independent* because the *measurements are from different people*. Knowing something about the distribution of values in the first sample does not inform you about the distribution of values in the second.

In order to compare the mean values of a quantitative variable between two groups:

- Apply the *paired t-test* when the two samples are *dependent*.
- Apply the *two-sample t-test* when the two samples are *independent*.

Stat Tool 3.4 Two Variances Test

Considering a *quantitative variable Y*, use the *two variances test* to determine whether two population variances σ_1^2, σ_2^2, (or standard deviations, which are the square roots of variances) are equal.

The two variances test requires two *independent random samples* selected from two populations (Stat Tool 1.2, 3.3). We use the *sample variances* to draw conclusions about the *population variances*.

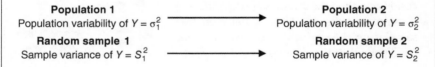

Population 1
Population variability of $Y = \sigma_1^2$ ⟶ **Population 2**
Population variability of $Y = \sigma_2^2$

Random sample 1
Sample variance of $Y = S_1^2$ ⟶ **Random sample 2**
Sample variance of $Y = S_2^2$

➤ *Example 3.4.* It is of interest to compare the *variability* (e.g. the variances) of blood pressure between a test drug and the placebo.

The *two variances test* considers the *sample variances* in two *independent* random samples:

It evaluates the *ratio* between the two *sample variances*:

and allows us to draw conclusions about the *ratio* between *population variances*:

1. Test drug $S_{1(Drug)}^2 = 12$

2. Placebo $S_{2(Placebo)}^2 = 25$

S_1^2 / S_2^2
⬇
σ_1^2 / σ_2^2

Stat Tool 3.4 (Continued)

by determining whether this ratio is:

$$\sigma_1^2/\sigma_2^2 \neq 1 \quad \sigma_1^2 \neq \sigma_2^2$$

- Statistically different from 1: the two population $\sigma_1^2/\sigma_2^2 = 1 \quad \sigma_1^2 = \sigma_2^2$ variances differ from one another.
- NOT statistically different from 1: the two population variances do not differ from one another.

The two variances test results in Minitab include both a confidence interval and a p-value:

- Use the confidence interval (usually a 95% or a 99% CI) to obtain a range of reasonable values for the ratio of population variances: σ_1^2 / σ_2^2.
- Use the p-value to determine whether the ratio of population variances differs from 1 (the two populations' variances are not equal).

For a two variances test:

Null Hypothesis:	**Alternative Hypothesis:**
Population variances are EQUAL.	Population variances are NOT EQUAL.
$H_0: \sigma_1^2 = \sigma_2^2 \quad \left(or\ \sigma_1^2 / \sigma_2^2 = 1\right)$	$H_1: \sigma_1^2 \neq \sigma_2^2 \quad \left(or\ \sigma_1^2 / \sigma_2^2 \neq 1\right)$

Setting the significance level α usually at 0.05 or 0.01:

- If the p-value is less than α, reject the null hypothesis and conclude that the two variances are statistically different. — p-value $< \alpha$ — Reject the null hypothesis of equality of population variances.
- If the p-value is greater than or equal to α, fail to reject the null hypothesis and conclude that the two variances are NOT statistically different. — p-value $\geq \alpha$ — Fail to reject the null hypothesis of equality of population variances.

If you *reject the null hypothesis* of equality of variances, remember to check the *practical significance* of the result (Stat Tool 1.16).

3.3.2 Comparing Means Between Two Groups

To determine if there is any significant difference in the *mean* of thickness at $30 + 5$ mm from open end between the two formulations, let's proceed in the following way:

Step 1 – Perform a descriptive analysis (Stat Tool 1.3) of the variable "Open_ end" stratifying by formulations.

Step 2 – Assess the null and the alternative hypotheses (Stat Tool 1.15) and apply the two-sample t-test (Stat Tools 1.16, 3.5).

3.3.2.1 Step 1 – Perform a Descriptive Analysis of the Variable "Open_end" stratifying by Formulations

In step 1 of Section 3.3.1 we have already performed the descriptive analysis of variable "Open_end," stratifying by formulations and displaying the dotplots, the boxplots, and calculating the descriptive measures.

3.3.2.1.1 Interpret the Results of Step 1 We take back the considerations made previously in Section 3.3.1.1.1 about the distribution of the thickness at 30 + 5 mm from open end. In particular, we noticed that the mean was equal to 0.046 mm for group A and 0.048 for group B.

From a *descriptive point of view*, formulation B seems to have a slightly higher mean than A. The two means seem to differ. Is this difference statistically significant or NOT? Do the sample results lead to rejection of the null hypothesis of equality of the population means for the two formulations? Apply the two-sample t-test to answer these questions.

3.3.2.2 Step 2 – Assess the Null and the Alternative Hypotheses and Apply the Two-Sample t-Test

The appropriate procedure to determine whether the means of thickness measured at 30 + 5 mm from open end differ between the two formulations is the two-sample test, as the two random samples are independent (Stat Tools 3.3, 3.5). What are the *null* and *alternative hypotheses*?

Null Hypothesis

H_0: For the two formulations, the means of thickness at 30 + 5 mm from open end are equal.

Alternative Hypothesis

H_1: For the two formulations, the means of thickness at 30 + 5 mm from open end are not equal.

Look at the alternative hypothesis: the group A mean can be greater or less than the group B mean. This is a bidirectional alternative hypothesis. In this case the test is called a *two-tailed* (or *two-sided*) *test*.

Next, we'll look at the test results and decide whether to reject or fail to reject the null hypothesis.

To apply the two-sample t-test, go to:

Stat > Basic Statistics > 2 – sample t

In the drop-down menu at the top, choose **Both samples are in one column**. In **Samples**, select the variable "Open_end" and in **Sample IDs**, select the variable "Formulation," then click **Options**. Under **Alternative hypothesis**, choose **Difference ≠ hypothesized difference**, to apply a two-sided test. Check the option **Assume equal variances** because in

Section 3.3.1 after applying the two variances test, we concluded in favor of the equality of variances between the two formulations. If in the application of the two variances test, you did not reject the null hypothesis, uncheck this option. Click **OK** in each dialog box.

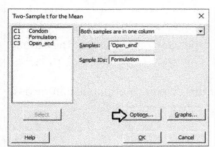 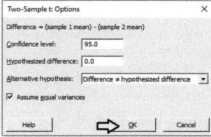

Stat Tool 3.5 Two-Sample t-Test

Considering a *quantitative variable Y*, use the *two-sample t-test* to determine whether two population means μ_1, μ_2, are equal.

The two-sample t-test requires two *independent random samples* selected from two populations (Stat Tool 1.2, 3.3). We use the *sample means* to draw conclusions about the *population means*.

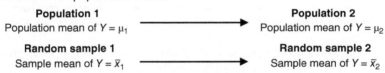

Population 1	**Population 2**
Population mean of $Y = \mu_1$	Population mean of $Y = \mu_2$

Random sample 1	**Random sample 2**
Sample mean of $Y = \bar{x}_1$	Sample mean of $Y = \bar{x}_2$

➢ *Example 3.5.* It is of interest to compare the blood pressure *mean* between a test drug and the placebo.

The *two-sample t-test* considers the *sample means* in two *independent* random samples:

It evaluates the *difference* between the two *sample means*:

and allows to draw conclusions about the *difference* between *population means*:

by determining whether this difference is:

$$\bar{x}_1 - \bar{x}_2$$

$$\Downarrow$$

$$\mu_1 - \mu_2$$

$$\mu_1 - \mu_2 \neq 0, \ \mu_1 \neq \mu_2$$
$$\mu_1 - \mu_2 = 0, \ \mu_1 = \mu_2$$

- Statistically different from 0: the two population means differ from one another.
- NOT statistically different from 0: the two population means do not differ from one another.

Stat Tool 3.5 (Continued)

The two-sample t-test results in Minitab include both a confidence interval and a p-value:

- Use the confidence interval (usually a 95% or a 99% CI) to obtain a range of reasonable values for the difference of population means: $\mu_1 - \mu_2$. The central point of the confidence interval is the difference of sample means: $\bar{x}_1 - \bar{x}_2$ (point estimate of $\mu_1 - \mu_2$).
- Use the p-value to determine whether the difference of population means differs from 0 (i.e. the two populations' means are not equal).

For a two-sample t-test:

Null Hypothesis:	**Alternative Hypothesis:**
Population means are EQUAL.	Population means are NOT EQUAL.
$H_0: \mu_1 = \mu_2$ (*or* $\mu_1 - \mu_2 = 0$)	$H_1: \mu_1 \neq \mu_2$ (*or* $\mu_1 - \mu_2 \neq 0$)

Setting the significance level α usually at 0.05 or 0.01:

- If the p-value is less than α, reject the null hypothesis and conclude that the two means are statistically different. — p-value $< \alpha$ — Reject the null hypothesis of equality of population means.
- If the p-value is greater than or equal to α, fail to reject the null hypothesis and conclude that the two means are NOT statistically different. — p-value $\geq \alpha$ — Fail to reject the null hypothesis of equality of population means.

If you *reject the null hypothesis* of equality of means, remember to check the *practical significance* of the result (Stat Tool 1.16).

3.3.2.2.1 Interpret the Results of Step 2 The difference between the two sample means −0.002069, is a point estimate of the population mean difference. The confidence interval provides a range of likely values for the population mean difference. We can be 95% confident that on average the true difference is between −0.003769 and −0.002924. Notice that the CI does not include the value 0. Because the test's p-value (0.018) is less than the significance level $\alpha = 0.05$, we reject the null hypothesis. There is enough evidence that the mean of the thickness at $30+5\,mm$ from open end is different between the two formulations. The difference between the two means is *statistically significant*.

Two-Sample T-Test and CI: Open_end; Formulation

Method

μ_1: mean of Open_end when Formulation = A
μ_2: mean of Open_end when Formulation = B
Difference: $\mu_1 - \mu_2$
Equal variances are assumed for this analysis.

Descriptive Statistics: Open_end

Formulation	N	Mean	StDev	SE Mean
A	23	0.04596	0.00230	0.00048
B	25	0.04803	0.00340	0.00068

Estimation for Difference

Difference	Pooled StDev	95% CI for Difference
−0.002069	0.002924	(−0.003769; −0.000368)

Test

Null hypothesis	$H_0: \mu_1 - \mu_2 = 0$
Alternative hypothesis	$H_1: \mu_1 - \mu_2 \neq 0$

t-Value	DF	p-Value
−2.45	46	0.018

3.3.3 Comparing Two Proportions

To determine whether the *proportion* of condoms with thickness less than or equal to 0.045 mm differs between the two formulations, let's proceed in the following way:

Step 1 - For the two formulations, calculate the sample proportion of condoms with thickness less than or equal to 0.045 mm.

Step 2 - Assess the *null* and the *alternative hypotheses* (Stat Tool 1.15) and apply the *two proportions test* (Stat Tools 1.16, 3.6).

3.3.3.1 Step 1 – Calculate the Sample Proportions of Condoms with Thickness Less than or Equal to 0.045 mm

Let us create a new categorical variable from variable "Open_end" that assumes the two categories: "thickness less than or equal to 0.045," "thickness greater than 0.045."

To create a new categorical variable, go to: **Data > Recode > To Text** Under **Recode values in the following columns**, select the variable "Open_end." Then, under **Method**, choose the option **Recode ranges of values**. Define the lower and upper endpoint of each category and in **Endpoints to include**, specify **Lower endpoint only**. Then click **OK** in the dialog box.

Note that the new variable "Recoded Open_end" will appear at the end of the current worksheet. This variable is a categorical variable assuming two categories: "Less than or equal to 0.045," "Greater than 0.045."

To calculate the sample proportion of condoms with thickness less than or equal to 0.045 mm by formulations, go to:

Stat > Tables > Cross Tabulation and Chi – Square

In the drop-down menu at the top, choose **Raw data (Categorical variables)**. In **Rows,** select the variable "Recoded Open_end" and in **Columns,** choose "Formulation." Then, under **Display,** check the options **Count** and **Column Percents.** Click **OK** in the dialog box.

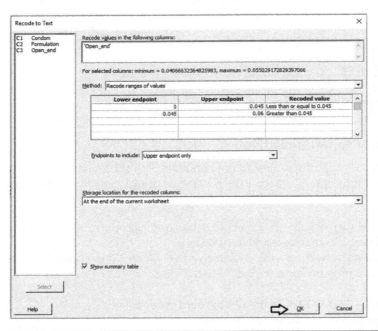

↓	C1	C2-T	C3	C4-T	C5	C6
	Condom	Formulation	Open_end	Recoded Open_end		
1	1	A	0.049	Greater than 0.045		
2	2	A	0.049	Greater than 0.045		
3	3	A	0.048	Greater than 0.045		
4	4	A	0.048	Greater than 0.045		
5	5	A	0.044	Less than or equal to 0.045		
6	6	A	0.045	Greater than 0.045		

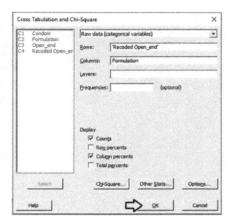

To graphically display the sample proportions of condoms with thickness less than or equal to 0.045 mm, go to:

Graph > Pie Chart

In **Categorical variables**, select the variable "Recoded Open_end." Click **Labels** and in the dialog box select **Slice Labels**. Check the option **Percent** and click **OK**.

In the main dialog box select **Multiple Graphs**. In the next window, select **By variables**. In the new dialog box under **By variables with groups on same graph**, select "Formulation." Click **OK** in each dialog box.

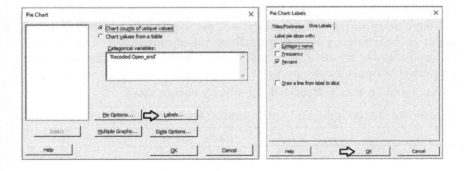

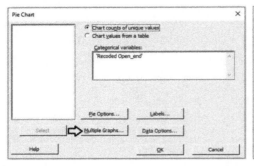

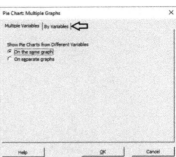

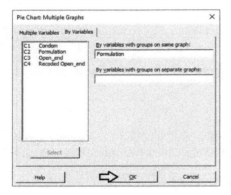

3.3.3.1.1 *Interpret the Results of Step 1*

The table stored in the Session window and the pie charts shows the frequency distribution of the variable "Recoded Open_end" by formulation. For group A, in 8 out of 23 condoms (sample percentage = 34.8%; sample proportion = 0.348), the thickness is less than or equal to 0.045 mm; for formulation B, this sample percentage is 20.0%.

From a *descriptive point of view*, the two percentages seem to be a little different. Is the difference of 14.8% statistically significant or NOT? Do the sample results lead to rejection of the null hypothesis of equality of the population proportions? Apply the two proportions test to answer these questions.

Tabulated Statistics: Recoded Open_end; Formulation

Rows: Recoded Open_end   Columns: Formulation

	A	B	All
Less than or equal to 0.045	8	5	13
	34.78	20.00	27.08
Greater than 0.045	15	20	35
	65.22	80.00	72.92

	A	B	All
All	23	25	48
	100.00	100.00	100.00

Cell Contents

Count

% of Column

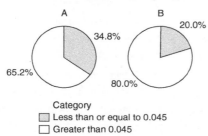

Pie Chart of Recoded Open_end

A   34.8%   B   20.0%

65.2%   80.0%

Category
☐ Less than or equal to 0.045
☐ Greater than 0.045

3.3.3.2 Step 2 – Assess the Null and the Alternative Hypotheses and Apply the Two Proportions Test

The appropriate procedure to determine whether the proportions of condoms with thickness less than or equal to 0.045 mm is different between the two formulations, is the two proportions test (Stat Tool 3.6). What are the *null* and *alternative hypotheses*?

Null hypothesis	H_0: For the two formulations, the proportions of condoms with thickness less than or equal to 0.045 mm are equal.
Alternative hypothesis	H_1: For the two formulations, the proportions of condoms with thickness less than or equal to 0.045 mm are not equal.

Look at the alternative hypothesis: The group A proportion can be greater or less than the group B proportion. This is a bidirectional alternative hypothesis. In this case, the test is called a *two-tailed* (or *two-sided*) *test*.

Next we'll look at the test results and decide whether to reject or fail to reject the null hypothesis.

 To apply the two proportions test, go to:

Stat > Basic Statistics > 2 Proportions

In the drop-down menu at the top, choose **Summarized data**. Under **Sample 1**, in **Number of events**, enter 8 (the number of condoms not exceeding 0.045 mm as thickness for formulation A) and in **Number of trials**, enter 23 (the sample size). Under **Sample 2**, in **Number of events**, enter 5 (the number of condoms not exceeding 0.045 mm as thickness for formulation B) and in **Number of trials**, enter 25 (the sample size). Click **Options**. Under **Alternative hypothesis**, choose **Difference ≠ hypothesized difference,** to apply a two-sided test, and under **Test Method** leave the default option **Estimate the proportions separately.** Click **OK** in each dialog box.

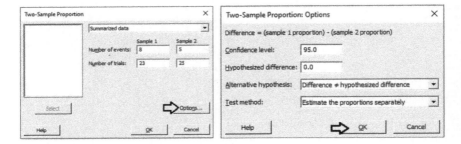

3.3.3.2.1 *Interpret the Results of Step 2*

The difference between the two-sample proportions 0.148 is a point estimate of the population difference. We can be 95% confident that on average the true difference is between −0.102119 and 0.397771. Notice that the CI includes the value 0.

Minitab uses the normal approximation method and Fisher's exact method to calculate the p-values for the two proportions test. If the number of events and the number of non-events is at least 5 in both samples, use the smaller of the two p-values. If either the number of events or the number of non-events is less than 5 in either sample, the normal approximation method may be inaccurate. Because the test's p-value (0.246) is greater than the significance level $\alpha = 0.05$, we fail to reject the null hypothesis. There is not enough evidence that the proportions of condoms with thickness less than or equal to 0.045 mm is different between the two formulations. The difference between the two proportions is *not statistically significant.*

Test and CI for Two Proportions Method
p_1: proportion where Sample 1 = Event
p_2: proportion where Sample 2 = Event
Difference: $p_1 - p_2$

Descriptive Statistics

Sample	N	Event	Sample p
Sample 1	23	8	0.347826
Sample 2	25	5	0.200000

Estimation for Difference

Difference	95% CI for Difference
0.147826	(−0.102119; 0.397771)

CI closed on normal approximation

Test

Null hypothesis	H_0: $p_1 - p_2 = 0$
Alternative hypothesis	H_1: $p_1 - p_2 \neq 0$

Method	z-Value	p-Value
Normal approximation	1.16	0.246
Fisher's exact		0.335

Stat Tool 3.6 Two Proportions Test

Suppose that the research interest is to compare the proportions of statistical units with a characteristic of interest between two groups. Use the *two proportions test* to determine whether two population proportions π_1, π_2, are equal.

The two proportions test requires two *independent random samples* selected from two populations (Stat Tools 1.2, 3.3). We use the *sample proportions* to draw conclusions about the *population proportions*.

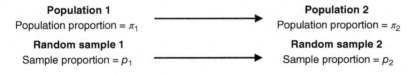

Population 1		**Population 2**
Population proportion = π_1	→	Population proportion = π_2
Random sample 1		**Random sample 2**
Sample proportion = p_1	→	Sample proportion = p_2

> *Example 3.6.* It is of interest to compare the *proportion* of patients with blood pressure (systolic) less than 120 mm Hg, between a test drug and the placebo.

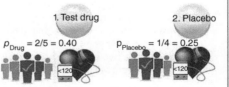

Stat Tool 3.6 (Continued)

The *two proportions test* considers the *sample proportions* in two *independent* random samples:

$$p_1 - p_2$$
$$\Downarrow$$
$$\pi_1 - \pi_2$$

It evaluates the *difference* between the two *sample proportions*:

and allows us to draw conclusions about the *difference* between *population proportions*:

$$\pi_1 - \pi_2 \neq 0, \quad \pi_1 \neq \pi_2$$
$$\pi_1 - \pi_2 = 0, \quad \pi_1 = \pi_2$$

by determining whether this difference is:

- Statistically different from 0: the two population proportions differ from one another.
- NOT statistically different from 0: the two population proportions do not differ from one another.

The two proportions test results in Minitab include both a confidence interval and a p-value:

- Use the confidence interval (usually a 95% or a 99% CI) to obtain a range of reasonable values for the difference of population proportions: $\pi_1 - \pi_2$. The central point of the confidence interval is the difference of sample proportions: $p_1 - p_2$ (point estimate of $\pi_1 - \pi_2$).
- Use the p-value to determine whether the difference of population proportions differs from 0 (i.e. the two populations' proportions are not equal).

For a two proportions test:

Null Hypothesis:	**Alternative Hypothesis:**
Population proportions are EQUAL.	Population proportions are NOT EQUAL.
$H_0: \pi_1 = \pi_2 \quad (or\ \pi_1 - \pi_2 = 0)$	$H_1: \pi_1 \neq \pi_2 \quad (or\ \pi_1 - \pi_2 \neq 0)$

Setting the significance level α usually at 0.05 or 0.01:

- If the p-value is less than α, reject the null hypothesis and conclude that the two proportions are statistically different.
- If the p-value is greater than or equal to α, fail to reject the null hypothesis and conclude that the two proportions are NOT statistically different.

p-value $< \alpha$ Reject the null hypothesis of equality of population proportions.

p-value $\geq \alpha$ Fail to reject the null hypothesis of equality of population proportions.

If you *reject the null hypothesis* of equality of proportions, remember to check the *practical significance* of the result (Stat Tool 1.16).

3.4   Case Study for Paired Data: Fragrance Project

The Fragrance Project aims to introduce an innovative shaving oil differentiated from existing shaving creams and foams. It is of interest to assess two fragrances (A and B) to investigate if there are any consumer perceived differences in their in-use characteristics and if so, which is the more suitable.

A total of 30 female respondents (panelists) are presented with the two fragrances and asked to answer the question, "How suitable or unsuitable is this fragrance for a shaving product?" and then assign an integer score from 0 (very unsuitable) to 10 (very suitable).

The research team needs to assess the following issue:

- Is there any significant difference in the *appropriateness* between the two fragrances?

The variables setting is the following:

- Variable "Fragrance" is a *nominal categorical variable,* assuming two different categories "A" and "B."
- Variable "Appropriateness" is a *discrete quantitative variable,* assuming values from 0 to 10.

Column	Variable	Type of data	Label
C1	Panelist		Panelists' code
C2-T	Fragrance	Categorical data	A = first fragrance; B = second fragrance
C3	Appropriateness	Numeric data	Appriopriateness of fragrance: 0 = very unsuitable to 10 = very suitable

File: Fragrance_Project.xlsx

The case study represents a typical *two-sample inferential problem on means,* but the scores of appropriateness for fragrance A and B are *dependent paired data* because each panelist tests both the fragrances at different times (Stat Tool 3.3). The sampling units are the 30 female respondents who assess the two fragrances and give opinions at two different times.

Use the *paired t-test* to compare a population mean between two *dependent* groups or samples of statistical units (Stat Tool 3.7).

In the Fragrance Project, to investigate if there are any consumer-perceived differences in appropriateness between the two fragrances, work out the statistical analysis in the following way:

Step 1 – Perform a descriptive analysis (Stat Tool 1.3) of the variable "Appropriateness" stratified by the categorical variable "Fragrance."

Step 2 – Calculate the differences between the score for fragrance A and the score for fragrance B for each panelist, creating a new variable "Difference_A_B" and perform a descriptive analysis of this variable.

Step 3 – Apply the paired t-test (Stat Tool 3.7).

3.4.1.1 Step 1 – Descriptive Analysis of "Appropriateness" Stratified by "Fragrance"

Let's start the descriptive analysis of the variable "Appropriateness" stratified by the categorical variable "Fragrance," by displaying the frequency distribution of the "Appropriateness" by "Fragrance" through graphs that help to emphasize trend, pattern, and other important aspects of data distributions. "Appropriateness" being a quantitative discrete variable, use the dotplot or the histogram to display the distribution of the data. Add the boxplot and calculate means and measures of variability to complete the analysis.

To display the histogram, go to: **Graph > Histogram**

Choose the graphical option **With Groups** to display one histogram of Appropriateness for each fragrance. In the next dialog box, select "Appropriateness" in **Graph variables** and "Fragrance" in **Categorical variables for grouping**. Then, the option **Multiple Graphs** gives access to additional options that help to display the histograms in separate panels (one for each fragrance) of the same graph.

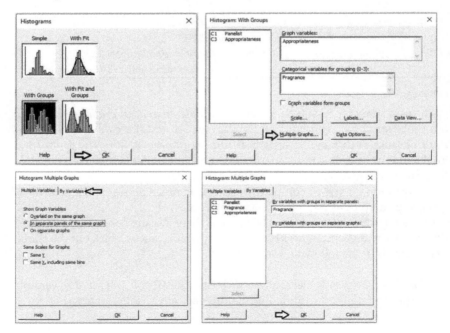

Use the tips in Chapter 2, section 2.2.2.1 to change the appearance of the histogram:

To display the Dotplot, go to: **Graph > Dotplot**

Choose the graphical option **With Groups** to display one dotplot of Appropriateness for each fragrance. In the next dialog box, select "Appropriateness" in **Graph variables** and "Fragrance" in **Categorical variables for grouping**.

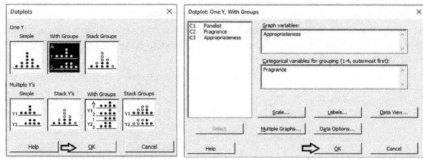

Use the tips previously referred to histograms to change the appearance of a dotplot.

To display the boxplot, go to: **Graph > Boxplot**

Choose the graphical option **With Groups** to display one boxplot of Appropriateness for each fragrance. In the next dialog box, select "Appropriateness" in **Graph variables** and "Fragrance" in **Categorical variables for grouping**.

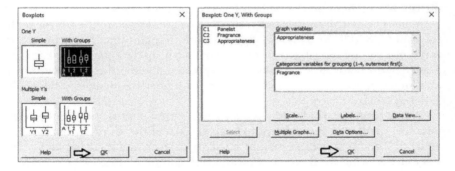

To change the appearance of a boxplot, use the tips previously referred to histograms, as well as the following ones:

- To display the boxplot horizontally: double-click on any scale value on the horizontal axis. A dialog box will open with several options to define. Select from these options: Transpose value and category scales.

- To add the mean value to the boxplot: right-click in the area inside the border of the boxplot, then select Add > Data display > Mean symbol.

To display the descriptive measures (means, measures of variability, etc.), go to:

Stat > Basic Statistics > Display Descriptive Statistics

In the dialog box, under **Variables** select "Appropriateness" and under **By variables** select "Fragrance," then click on **Statistics** to open a dialog box displaying a range of possible statistics to choose from. In addition to the default options, select **Coefficient of variation, Interquartile range** and **Range**.

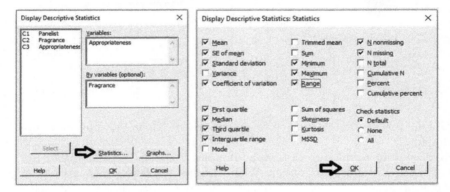

3.4.1.1.1 Interpret the Results of Step 1 From a descriptive point of view, let us compare the shape, central and non-central tendency and variability of the distribution of appropriateness between the two fragrances.

Shape. The distributions of appropriateness are skewed to the left for both fragrances: middle and high scores are more frequent than low scores. For fragrance A, the distribution is unimodal with mode equal to 8. For fragrance B, the distribution is reasonably spread out with no concentration on a unique value; scores 6, 9, and 10 show the higher frequencies (Stat Tools 1.4–1.5, 1.11).

Central tendency. For both fragrances, the central tendency of appropriateness (the median score) is equal to 7: about 50% of the scores is less than or equal to 7 and 50% greater than or equal to 7. Mean and median overlap for fragrance A, while for fragrance B, with a distribution more skewed to the left, mean and median are not close in value. The median score is a better indicator than the mean (Stat Tools 1.6, 1.11).

Non-central tendency. For fragrance A, 25% of the scores are less than 6 (first quartile Q_1) and 25% are greater than 8 (third quartile Q_3). For fragrance B, 25% of the scores are less than 4.75 (first quartile Q_1) and 25% are greater than 9 (third quartile Q_3) (Stat Tools 1.7, 1.11).

Variability. Observing the length of the boxplots (ranges) and the width of the boxes (Interquartile ranges, IQRs), fragrance B shows more variability. Its scores vary from 1 (minimum) to 10 (maximum). For fragrance A, no respondent gave a score less than 3 and the maximum observed difference between evaluations was 7 (Range). 50% of middle evaluations vary from 6 to 8 for fragrance A (IQR = maximum observed difference between evaluations equal to 2) and from 4.75 to 9 for fragrance B (IQR = 4.25) (Stat Tools 1.8, 1.11).

The average distance of the scores from the mean (standard deviation) is 1.93 for fragrance A (scores vary on average from 7 − 1.93 = 5.07 to 7 + 1.93 = 8.93) and 2.56 for fragrance B (scores vary on average from 6.77 − 2.56 = 4.21 to 6.77 + 2.56 = 9.33) (Stat Tools 1.9).

The coefficient of variation is less for fragrance A (CV = 27.57%) confirming the lower variability of the scores for this fragrance (Stat Tool 1.10).

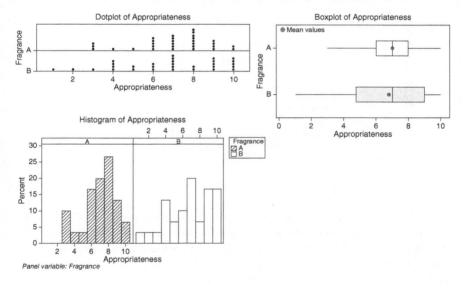

Statistics

Variable	Fragrance	N	N*	Mean	SE Mean	StDev	CoefVar	Minimum	Q1	Median
Appropriateness	A	30	0	7.000	0.352	1.930	27.57	3.000	6.000	7.000
	B	30	0	6.767	0.467	2.555	37.76	1.000	4.750	7.000

Variable	Fragrance	Q3	Maximum	Range	IQR
Appropriateness	A	8.000	10.000	7.000	2.000
	B	9.000	10.000	9.000	4.250

3.4.1.2 Step 2 – Descriptive Analysis of "Difference_A_B"

Let's calculate the differences between the score for fragrance A and the score for fragrance B for each panelist, creating a new variable "Difference_A_B," and perform a synthetic descriptive analysis to get a sense of the mean and variability in the differences between scores. "Difference_A_B" being a quantitative continuous variable, use the boxplot and calculate means and measures of variability to complete the analysis.

 To split the column "Appropriateness" into separate columns according to the values of "Fragrance,"
go to: **Data > Unstack columns**
In the dialog box, select "Appropriateness" in **Unstack the data in**, and "Fragrance" in **Using subscripts in**, then check the option **After last column in use**. The worksheet will include two new variables "Appropriateness_A" and "Appropriateness_B."

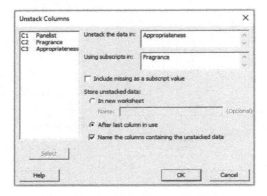

To calculate the differences between the new variables, go to: **Calc > Calculator**
In the dialog box, write the name of variable "Difference_A_B" in **Store result in variable**, and type the expression under **Expression**, selecting variables "Appropriateness_A" and "Appropriateness_B" from the list on the left and using the numeric keypad. The worksheet will include a new variable "Difference_A_B."

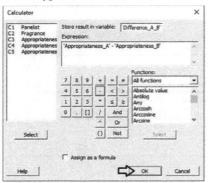

To display the boxplot, go to: **Graph** > **Boxplot**

Choose the graphical option **Simple** under **One Y**. In the next dialog box, select "Difference_A_B" in **Graph variables**.

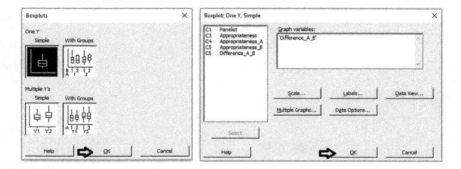

To display the descriptive measures (means, measures of variability, etc.), go to:

Stat > **Basic Statistics** > **Display Descriptive Statistics**

In the dialog box, under **Variables** select "Difference_A_B" and leave **By variables** blank, then click on **Statistics** to open a dialog box displaying a range of possible statistics to choose from. In addition to the default options, select **Interquartile range** and **Range** and uncheck **Coefficient of variation**.

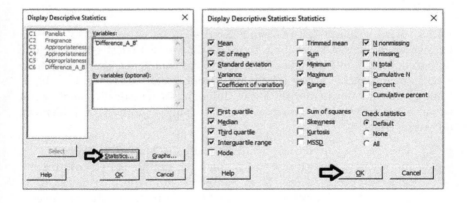

3.4.1.2.1 Interpret the Results of Step 2 We may see that on average the difference in the score of appropriateness of fragrances is 0.23 (median = 0) (Stat Tools 1.6, 1.11). 50% of middle differences vary from −1.25 to 2.25 (IQR = 3.5) (Stat Tools 1.8, 1.11). The average distance of the differences from their mean (standard deviation) is 0.44 (differences vary on average from 0.23−0.44 = −0.21 to 0.23 + 0.44 = 0.67) (Stat Tool 1.9).

Statistics

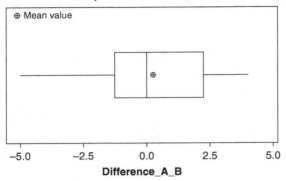

Boxplot of Difference_A_B

Variable	N	N*	Mean	SE Mean	StDev	Minimum	Q1	Median	Q3	Maximum	Range
Difference_A_B	30	0	0.233	0.444	2.431	−5.000	−1.250	0.000	2.250	4.000	9.000

Variable	IQR
Difference_A_B	3.500

From a descriptive point of view, the two fragrances show similar appropriateness, but we need to analyze the paired t-test results to determine whether the mean difference between the two fragrances is statistically different from 0.

3.4.1.3   Step 3 – Paired t-Test on Mean Difference

Let's apply the paired t-test (Stat Tool 3.7). The paired t-test results include both a confidence interval and a p-value. Use the confidence interval to obtain a range of reasonable values for the population mean difference. Use the p-value to determine whether the mean difference is statistically equal or not equal to 0. If the p-value is less than α, reject the null hypothesis that the mean difference equals 0.

To apply the paired t-test, go to: **Stat** > **Basics Statistics** > **Paired t**
In the dialog box, choose the option **Each sample is in a column** and select "Appropriateness_A" in **Sample 1** and "Appropriateness_B" in **Sample 2**. Click **Options** if you want to modify the significance level of the test or specify a directional alternative. Click on **Graphs** to display the boxplot with the null hypothesis and the confidence interval for the mean difference.

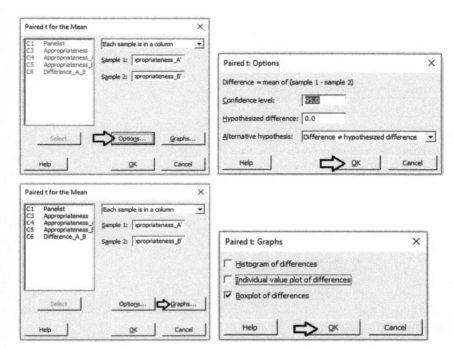

3.4.1.3.1 Interpret the Results of Step 3   The sample mean difference 0.233 is an estimate of the population mean difference. The confidence interval provides a range of likely values for the mean difference. We can be 95% confident that on average the true mean difference is between −0.674 and 1.141. Notice that the CI includes the value 0.

Because the p-value (0.603) is greater than 0.05, we fail to reject the null hypothesis. There is not enough evidence that the mean difference is different from 0. The mean difference of 0.233 between the appropriateness of fragrances A and B is not statistically significant. The two fragrances show similar appropriateness score means.

Paired T-Test and CI: Appropriateness_A; Appropriateness_B
   Descriptive Statistics

Sample	N	Mean	StDev	SE Mean
Appropriateness_A	30	7.000	1.930	0.352
Appropriateness_B	30	6.767	2.555	0.467

Estimation for Paired Difference

Mean	StDev	SE Mean	95% CI for μ_difference
0.233	2.431	0.444	(−0.674; 1.141)

μ_difference: mean of (Appropriateness_A − Appropriateness_B)

Test

Null hypothesis	H_0: μ_difference = 0
Alternative hypothesis	H_1: μ_difference ≠ 0

t-Value	p-Value
0.53	0.603

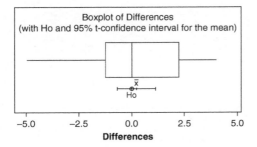

Boxplot of Differences
(with Ho and 95% t-confidence interval for the mean)

Differences

Stat Tool 3.7 Paired t-Test

Considering a quantitative variable, use the *paired t-test* to determine whether two population means are equal when data are dependent (paired) rather than independent (Stat Tool 3.3).

The paired t-test requires two dependent random samples (Stat Tool 1.2) selected from the two populations.

Paired or dependent data usually arise when we observe or measure categorical or numerical variables on the same individual or statistical unit at different points in time. In the two dependent random samples, an observation in one sample corresponds to an observation in the other.

> *Example 3.7.* A clinical trial might record the blood pressure in the same group of patients *before* and *after* taking a new drug. In this case, data are *dependent* or *paired*, as it is likely the blood pressure after taking the new drug will be related to the blood pressure of that patient before the drug was given.

Before Without drug	After With drug	Difference (Before – After)
120	118	120 − 118 = 2
116	113	116 − 113 = 3

The paired t-test considers the *differences* between a value in the first sample and its corresponding value in the second sample:

Stat Tool 3.7 (Continued)

It evaluates the *mean of all differences* between the first sample and second sample:

and allows us to draw conclusions about the *population mean difference*:

by determining whether this mean is:

$$\bar{x}_{DIFF}$$
$$\Downarrow$$
$$\mu_{DIFF}$$

$$\mu_{DIFF} \neq 0$$
$$\mu_{DIFF} = 0$$

- Statistically different from 0: There is a difference between the two populations or groups (without drug / with drug).
- NOT statistically different from 0: There is no difference between the two populations.

The paired t-test results in Minitab include both a confidence interval and a p-value:

- Use the confidence interval (usually a 95% or a 99% CI) to obtain a range of reasonable values for the population mean difference: μ_{DIFF}. The central point of the confidence interval is the sample mean difference: $\bar{x}_{DIFF}$ (point estimate of μ_{DIFF}).
- Use the p-value to determine whether the population mean difference differs from 0.

For a paired t-test:

Null Hypothesis:	**Alternative Hypothesis:**
Population means are EQUAL.	Population means are NOT EQUAL.
H_0: $\mu_{DIFF} = 0$	H_1: $\mu_{DIFF} \neq 0$

Setting the significance level α usually at 0.05 or 0.01:

- If the p-value is less than α, reject the null hypothesis and conclude that the mean difference is statistically different from 0.
  p-value $< \alpha$ — Reject the null hypothesis that population mean difference equals 0.

- If the p-value is greater than or equal to α, fail to reject the null hypothesis and conclude that the mean difference is NOT statistically different from 0.
  p-value $\geq \alpha$ — Fail to reject the null hypothesis that population mean difference equals 0.

If you *reject the null hypothesis*, remember to check the *practical significance* of the result (Stat Tool 1.16).

3.5 Case Study: Stain Removal Project

A Stain Removal Project aims to complete the development of an innovative dishwashing cleaning product. The research team needs to identify which formula from four component concentrations, maximizes stain release on several stains. Different *combinations of component concentrations* will be tested on a random sample of fabric specimens and reflectance measures (scores varying from 1: very bad performance, to 100: very good performance) will be obtained by a spectrophotometer. We need to plan and analyze an experimental study in order to identify which formula could maximize the cleaning performance.

3.5.1 Plan of the General Factorial Experiment

During a previous screening phase (for a review on screening designs, see Chapter 2), investigators selected four components affecting cleaning performance by using a two-level factorial design. We will denote these four components as A, B, C, D. After identifying the key input factors, the experimental work can proceed with a *general factorial design* increasing the number of levels to optimize responses (Stat Tool 2.2). The research team will test the following concentrations of the four components:

Components (Factors)	A	B	C	D
Concentrations (Levels)	10	0	0	0
	25	0.15	1.5	5
	40	0.3	3.0	10

For the present study:

- The *four components* represent the *key factors.*
- Each factor has three *levels* (*concentrations*).
- The *cleaning performance scores* (reflectance measures) for a list of stains represent *the response variables* (one for each stain).

For logistical reasons, investigators need to arrange the experiment in such a way that only 40 different combinations of factor levels will be tested.

We will proceed initially by constructing a *full* factorial design (all combinations of factor levels are tested) and then selecting an *optimal* design (only a *suitable subset* of all combinations are tested) in order to reduce the number of runs, thus respecting the restriction to 40 runs. Furthermore, we will reasonably assume that interactions among more than two factors are negligible.

An *optimal* design (Johnson, R. T., Montgomery, D. C., Jones, B. A., 2011) is a design that is *best* in relation to a particular statistical criterion (Anderson, M. J. and Whitcomb, P. J., 2014). Computer programs are required to construct such designs.

Remember that when feasible, the experiment should include at least two replicates of the design, i.e. the researcher should run each factor level combination at least twice. Replicates are multiple independent executions of the same experimental conditions (Stat Tool 2.5). In the present case study we will consider two *replicates* of each run.

To create the design for our factorial experiment, let's proceed in the following way:

Step 1 – Create a full factorial design.
Step 2 – Reduce the full design to an optimal design.
Step 3 – Assign the designed factor level combinations to the experimental units and collect data for the response variable.

3.5.1.1 Step 1 – Create a Full Factorial Design

Let's begin to create a full factorial design, considering three levels for each factor.

To create a full factorial design, go to: **Stat > DOE > Factorial > Create Factorial Design**

Choose **General full factorial design** and in **Number of factors** specify 4. Click on **Designs**. In the next dialog box, enter the names of the factors (A to D) and the number of levels. Under **Number of replicates**, specify **1**: We will request two replicates in step 2. Click **OK** and in the main dialog box select **Factors**. Select **Numeric** in **Type** as they are quantitative variables and specify the levels for each factor. If you have a categorical variable, choose **Text** in **Type**.

Proceed by clicking **Options** in the main dialog box and then uncheck the option **Randomize runs**: we will request the randomization in step 2 (Stat Tool 2.3). Click **OK** in the main dialog box and Minitab shows the design of the experiment in the worksheet (only the first five runs are shown here as an example), adding some extra columns that are useful for the statistical analysis. Information on the design is shown in the Session window. Note that the full factorial design includes 81 runs, which are all possible combinations of factor levels.

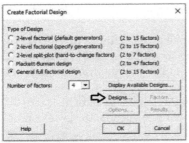

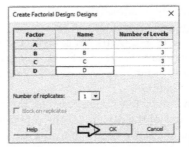

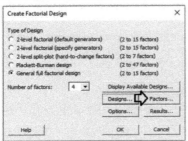

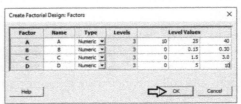

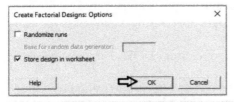

Multilevel Factorial Design
Design Summary

Factors:	4	Replicates:	1
Base runs:	81	Total runs:	81
Base blocks:	1	Total blocks:	1

Number of levels: 3; 3; 3; 3

3.5.1.2 Step 2 – Reduce the Full Factorial Design to an Optimal Design

Investigators need to arrange the experiment in such a way that only 40 different combinations of factor levels will be tested. We proceed by reducing the previous full factorial design to obtain an optimal design with 40 runs.

To create an optimal factorial design, go to:
Stat > DOE > Factorial > Select Optimal Design
In **Number of points in optimal design**, specify 40. Click on **Terms**. In the next dialog box, under **Include terms in the model up through order**, specify 2, as we want to consider only the interactions between two factors. Click **OK** in each dialog box and Minitab shows the optimal design in a new worksheet (only the first five runs are shown here as an example), as well as some information in the session panel. Note that in the Session window you can check which of the 81 runs have been selected in the optimal design.

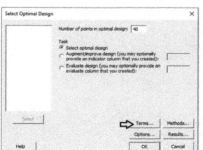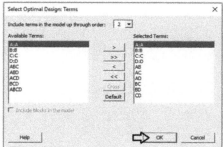

↓	C1	C2	C3	C4	C5	C6	C7	C8	C9
	StdOrder	RunOrder	PtType	Blocks	A	B	C	D	
1	27	27	1	1	10	0.30	3.0	10	
2	63	63	1	1	40	0.00	3.0	10	
3	75	75	1	1	40	0.30	0.0	10	
4	52	52	1	1	25	0.30	3.0	0	
5	70	70	1	1	40	0.15	3.0	0	

Optimal Design: A; B; C; D
Factorial design selected according to D-optimality
Number of candidate design points: 81
Number of design points in optimal design: 40
Model terms: A; B; C; D; AB; AC; AD; BC; BD; CD
Initial design generated by sequential method
Initial design improved by exchange method
Number of design points exchanged is 1

Optimal Design
Row number of selected design points: 27; 63; 75; 52; 70; 76; 45; 51; 69; 80; 3; 8; 20; 56; 40; 15; 16; 22; 33; 34; 39; 44; 46; 50; 58; 64; 68; 6; 23; 1; 11; 29; 54; 72; 79; 7; 17; 32; 21; 78

To request two replicates of each run, go to:

Stat > DOE > Modify Design

Under **Modification**, choose **Replicate design**, then click on **Specify**. In the next dialog box, under **Number of replicates to add**, specify **1**. Click **OK** in each dialog box and Minitab shows the replicated optimal design in the previous worksheet. The second replicates for each run will have the value **2** under the column **Block**.

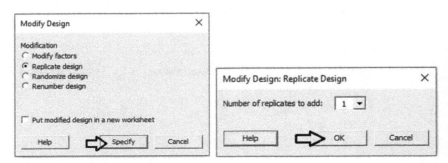

Before proceeding with the randomization, renumber the design by going to:

Stat > DOE > Modify Design

Under **Modification**, choose **Renumber design**, then click on **Put the modified design in a new worksheet**. Click **OK** in each dialog box and Minitab shows the replicated optimal design in a new worksheet.

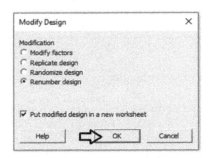

Let's complete the creation of the optimal design with the randomization of the runs. Go to:

Stat > DOE > Modify Design

Under **Modification**, choose **Randomize runs**, then click on **Specify**. In the next dialog box, select **Randomize entire design**. Click **OK** in each dialog box and Minitab shows the randomized optimal design in the previous worksheet (only the first five runs are shown here as an example).

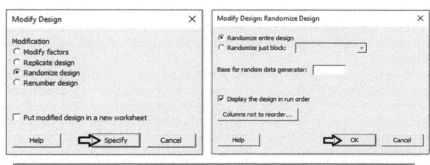

	C1	C2	C3	C4	C5	C6	C7	C8	C9
	StdOrder	RunOrder	PtType	Blocks	A	B	C	D	
1	155	1	1	2	40	0.30	0.0	10	
2	158	2	1	2	40	0.30	1.5	10	
3	86	3	1	2	10	0.00	1.5	10	
4	81	4	1	2	10	0.00	0.0	0	
5	101	5	1	2	10	0.30	0.0	10	

3.5.1.3 Step 3 – Assign the Designed Factor Level Combinations to the Experimental Units and Collect Data for the Response Variable

Collect the data for the response variables (reflectance measures) in the order set out in the column **RunOrder** in the optimal design. For example, the first formulation to test will be that with component A at level 40, component B at level 0.30, C at 0 and D at 10. Each formulation will be tested on two fabric specimens given we are considering two *replicates*. Note that the column "Blocks" has two values: 1 for the first replicate and 2 for the second.

Using a spectrophotometer, reflectance data (cleaning performance scores) will be measured in relation to several stains. To take into account the inherent variability in the measurement system, the cleaning performance for each stain will be measured at two different points on each fabric specimen. Remember that the individual measurements for each point on the piece of fabric are not replicates but *repeated measurements* because the two positions will be processed together in the experiment, receiving the same formulation simultaneously. Their *average* will be the response variable to analyze (Stat Tool 2.5).

Once all the designed factor level combinations have been tested, enter the recorded response values (cleaning performance scores) onto the worksheet containing the design. Now you are ready to proceed with the statistical analysis of the collected data.

3.5.2   Plan of the Statistical Analyses

The main focus of a statistical analysis of a *general factorial design* (Stat Tool 2.2) is to study in depth how several factors contribute to the explanation of a response variable and understand how the factors interact with each other. In a mature or reasonably well-understood system, a second important objective is the *optimization* of the response, i.e. the search for those combinations of factor levels that are linked to desirable values of response variables.

The appropriate procedures to use for these purposes is the *analysis of variance* (ANOVA) (Stat Tool 2.7) and, when at least one factor is a quantitative variable, a *response surface model*, can help investigators to find a suitable approximation for the unknown functional relationship between factors and response variables. This is useful to identify regions of factor levels that optimize response variables.

If your general factorial design includes all categorical factors, then you can analyze it as shown in Chapter 2, section 2.2.2. If at least one factor is a quantitative variable as in the Stain Removal Project, proceed in the following way:

Step 1 – Perform a descriptive analysis (Stat Tool 1.3) of the response variables.
Step 2 – Fit a response surface model to estimate the effects and determine the significant ones.
Step 3 – If need be, reduce the model to include the significant terms.
Step 4 – Optimize the responses.
Step 5 – Examine the shape of the response surface and locate the optimum.

For the Stain Removal Project, the dataset (File: Stain_Removal_Project.xlsx) opened in Minitab and containing two replicates for each formulation, i.e. for each combination of factor levels, and two repeated measurement for reflectance data on two stains (mustard and curry), appears as below.

	C1	C2	C3	C4	C5	C6	C7	C8	C9	C10	C11	C12	C13
	StdOrder	RunOrder	PtType	Blocks	A	B	C	D	Mustard 1	Mustard 2	Curry 1	Curry 2	
1	155	1	1	2	40	0.30	0.0	10	82.805	83.760	84.650	84.175	
2	158	2	1	2	40	0.30	1.5	10	84.325	84.720	84.700	85.400	
3	86	3	1	2	10	0.00	1.5	10	79.605	79.250	80.790	81.150	
4	81	4	1	2	10	0.00	0.0	0	77.565	78.330	78.545	78.740	
5	101	5	1	2	10	0.30	0.0	10	81.240	82.245	82.370	81.995	
...	...	...	...	...	...	...	...	...	...	...	...	...	
70	8	70	1	1	10	0.00	3.0	5	80.000	81.245	80.105	80.975	
71	44	71	1	1	25	0.15	3.0	5	82.415	82.785	82.815	83.150	
72	70	72	1	1	40	0.15	3.0	0	83.320	82.575	83.335	83.675	
73	75	73	1	1	40	0.30	0.0	10	83.475	83.840	85.760	85.915	
74	56	74	1	1	40	0.00	0.0	5	82.310	80.455	82.225	82.125	
...	...	...	...	...	...	...	...	...	...	...	...	...	

For example, records 1 and 73 represent two *replicates* of the formulation with component A at level 40, component B at level 0.30, C at 0 and D at 10. Columns C9 "Mustard 1" and C10 "Mustard 2" include the two *repeated measurements* for reflectance at two different positions on each fabric.

The variables setting is the following:

- Columns C1 to C4 are related to the previous creation of the optimal factorial design.
- Variables A, B, C, D are the quantitative factors, each assuming three levels.
- Variables from "Mustard 1" to "Curry 2" are *quantitative variables*, assuming values from 0 to 100.

Column	Variable	Type of data	Label
C5	A	Numeric data	Component A of the formulation with levels: 10, 25, 40
C6	B	Numeric data	Component B of the formulation with levels: 0, 0.15, 0.30
C7	C	Numeric data	Component C of the formulation with levels: 0, 1.5, 3
C8	D	Numeric data	Component D of the formulation with levels: 0, 5, 10
C9	Mustard 1	Numeric data	Reflectance for mustard, first measurement from 0 to 100
C10	Mustard 2	Numeric data	Reflectance for mustard, second measurement from 0 to 100
C11	Curry 1	Numeric data	Reflectance for curry, first measurement from 0 to 100
C12	Curry 2	Numeric data	Reflectance for curry, second measurement from 0 to 100

Stain_Removal_Project.xlsx

3.5.2.1 Step 1 – Perform a Descriptive Analysis of the Response Variables

Firstly calculate the average of the two repeated measurements for each stain, creating two new variables that will become the response variables to analyze.

 To calculate the average of the two repeated measurements for each stain, go to: **Calc > Calculator**

In the dialog box, write the name of a new column "Mustard" in **Store result in variable**, and type the expression under **Expression**, selecting variables "Mustard 1" and "Mustard 2" from the list on the left and using the numeric keypad. The worksheet will include the new response variable "Mustard". Do the same to create the new response variable "Curry."

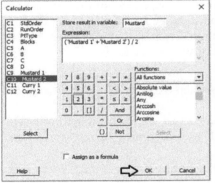

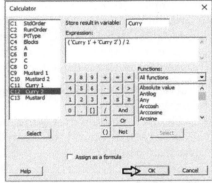

	C1	C2	C3	C4	C5	C6	C7	C8	C9	C10	C11	C12	C13	C14	C15
	StdOrder	RunOrder	PtType	Blocks	A	B	C	D	Mustard 1	Mustard 2	Curry 1	Curry 2	Mustard	Curry	
1	155	1	1	2	40	0.30	0.0	10	82.805	83.760	84.650	84.175	83.2825	84.4125	
2	158	2	1	2	40	0.30	1.5	10	84.325	84.720	84.700	85.400	84.5225	85.0500	
3	86	3	1	2	10	0.00	1.5	10	79.605	79.250	80.790	81.150	79.4275	80.9700	
4	81	4	1	2	10	0.00	0.0	0	77.565	78.330	78.545	78.740	77.9475	78.6425	
5	101	5	1	2	10	0.30	0.0	10	81.240	82.245	82.370	81.995	81.7425	82.1825	
...	...	...	...	...	...	...	...	...	...	...	...	...	...	...	...

The responses are quantitative variables, so let's use a boxplot to represent how the cleaning performance scores occurred in our sample, and calculate means and measures of variability to complete the descriptive analysis of each response.

To display the boxplot, go to: **Graph > Boxplot**

As the two responses have the same measurement scale of 0 to 100, the two boxplots can be displayed in the same graph. To do this, under **Multiple Y's**, check the graphical option **Simple** and in the next dialog box select "Mustard" and "Curry" in **Graph variables**. Then click **OK** in the dialog box.

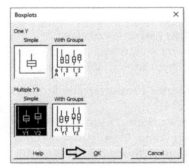

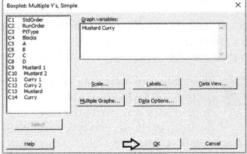

To change the appearance of a boxplot, use the tips in Chapter 2, Section 2.2.2.1 in relation to histograms, as well as the following ones:

- To display the boxplot horizontally: double-click on any scale value on the horizontal axis. A dialog box will open with several options to define. Select from these options: Transpose value and category scales.
- To add the mean value to the boxplot: right-click on the area inside the border of the boxplot, then select Add > Data display > Mean symbol.

To display the descriptive measures (means, measures of variability, etc.), go to:

Stat > Basic Statistics > Display Descriptive Statistics

In the dialog box, under **Variables** select "Mustard" and "Curry," then click on **Statistics** to open a dialog box displaying a range of possible statistics to choose from. In addition to the default options, select **Interquartile range**, **Range,** and **Coefficient of variation**.

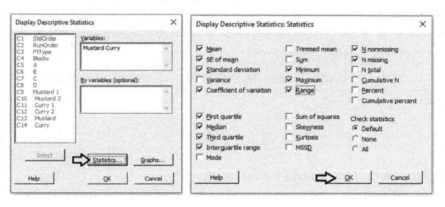

3.5.2.1.1 *Interpret the Results of Step 1*

Reflectance data is fairly symmetric for the curry stain and slightly skewed to the left with the presence of an outlier for the mustard stain (Stat Tools 1.4–1.5).

Regarding the *central tendency*, the mean and median overlap for cleaning performances on curry and mustard stains, with a higher mean value for curry (Stat Tools 1.6, 1.11).

With respect to *variability*, observing the length of the boxplots (ranges) and the width of the boxes (interquartile ranges, IQRs), the performance shows moderate spread for both stains. The average distance of the scores from the mean (standard deviation) is about 1.7/1.8, respectively, for Mustard and Curry (Stat Tools 1.9). Additionally, the coefficient of variation is about 2.1%/2.2% respectively for Mustard and Curry, confirming the similar spread of the cleaning scores for the two stains (Stat Tool 1.10).

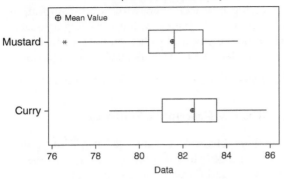

Boxplot of Mustard; Curry

Statistics

Variable	N	N*	Mean	SE Mean	StDev	CoefVar	Minimum	Q1	Median	Q3	Maximum
Mustard	80	0	81.509	0.187	1.676	2.06	76.567	80.426	81.594	82.926	84.523
Curry	80	0	82.403	0.196	1.754	2.13	78.610	81.043	82.535	83.548	85.837

Variable	Range	IQR
Mustard	7.955	2.500
Curry	7.227	2.505

3.5.2.2 Step 2 – Fit a Response Surface Model

When you have at least one quantitative factor in your general factorial design, you can analyze it by fitting a response surface model. We have added two new variables, "Mustard" and "Curry," to the worksheet including the factorial design. Therefore, before analyzing it, we must specify our factors.

 To define the response surface design, go to:

**Stat > DOE > Response Surface
> Define Custom Response Surface Design**

Under **Continuous Factors**, select the components A, B, C, D. If some of your factors are categorical, these must be specified under **Categorical Factors**. Click on **Low/High**. For each factor in the list, verify that the high and low settings are correct and if these levels are in the original units of the data, check the option **Uncoded** under **Worksheet Data Are**. Click **OK** in each dialog box. Now you are ready to analyze the response surface design.

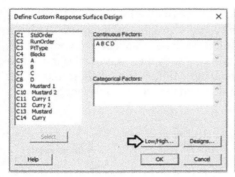
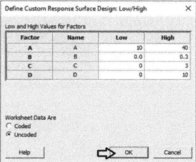

Note that in our example we don't have blocking factors (Stat Tool 2.4). Should there be any, in the main dialog box click on **Designs**. At the bottom of the screen, under Blocks, check the option **Specify by column**, and in the available space select the column representing the blocking variable.

Click **OK** in each dialog box. You would now be ready to analyze the response surface design.

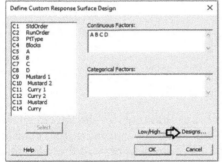
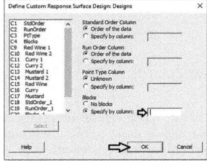

To analyze the response surface design, go to:

Stat > DOE > Response Surface > Analyze Response Surface Design

Select the first response variable "Mustard" under **Responses** and click on **Terms**. At the top of the next dialog box, in **Include the following terms**, specify **Full quadratic** to study all main effects (both linear and quadratic effects) and the two-way interactions between factors. As stated previously, we don't have blocking factors in our example (Stat Tool 2.4). If present, the grayed-out **Include blocks in the model option** would be selectable to consider blocks in the analysis. Click **OK** and now choose

Graphs in the main dialog box. Under **Effects plots** choose **Pareto** and **Normal**. These graphs help identify the terms (linear and quadratic factor effects and/or interactions) that influence the response and compare the relative magnitude of the effects, along with their statistical significance. Click **OK** in the main dialog box and Minitab shows the results of the analysis, both in the Session window and through the relevant graphs.

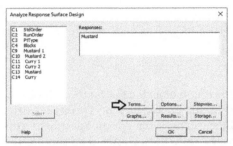
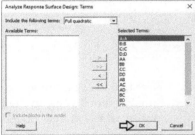

3.5.2.2.1 *Interpret the Results of Step 2*

Let us first consider the **Pareto chart** that shows which terms contribute the most to explaining the clinical performance on the mustard stain. Any bar extending beyond the reference line (considering a significance level equal to 5%) is related to a significant effect.

In our example, the main effects of three components B, C and D and their interactions are statistically significant at the 0.05 level. Also, the quadratic effect of factor B is marked as statistically significant.

From the Pareto chart, you can detect which effects are statistically significant, but you have no information on how these effects affect the response.

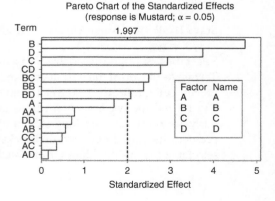

Use the **normal plot of the standardized effects** to evaluate the direction of the effects.

In the **normal plot**, the line represents the situation in which all the effects (main effects and interactions) are 0. Effects that are further from 0 and from the line are statistically significant. Minitab shows statistically significant and nonsignificant effects by giving different colors and shapes to the points. In addition, the plot indicates the direction of the effect.

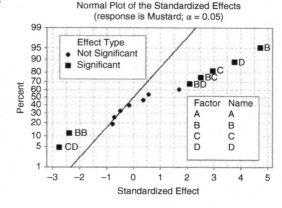

Positive effects (displayed on the right of the graph) increase the response when the factor moves from its low value to its high value.

Negative effects (displayed on the left of the line) decrease the response when moving from the factor's low value to its high value.

In our example, the main effects for B, C, D are statistically significant at the 0.05 level.

The cleaning scores seem to increase with higher levels of each component. Furthermore, the interactions between factor pairs are marked as statistically significant, but the interaction between components C and D lies on the left of the graph along with the quadratic effect of component B. We will clarify the meaning of this later.

In the ANOVA table we examine the p-values to determine whether any factors or interactions, or the blocks if present, are statistically significant. Remember that the p-value is a probability that measures the evidence against the null hypothesis. Lower probabilities provide stronger evidence against the null hypothesis.

For the main effects, the null hypothesis is that there is no significant difference in the mean performance score across levels of each factor. As we estimate a full quadratic model, we have two tests for each factor that analyze its linear and quadratic effect on the response.

For the two-factor interactions, H_0 states that the relationship between a factor and the response does not depend on the other factor in the term.

Usually we consider a significance level alpha equal to 0.05, but in an exploratory phase of the analysis, we may also consider a significance level equal to 0.10.

When the p-value is greater than or equal to alpha, we fail to reject the null hypothesis. When it is less than alpha, we reject the null hypothesis and claim statistical significance.

So, setting the significance level alpha to 0.05, which terms in the model are significant in our example? The answer is: the linear effects of B, C, D;

the two-way interactions among these three components and the quadratic effect of B. Now, you may want to reduce the model to include only significant terms.

When you use statistical significance to decide which terms to keep in a model, it is usually advisable not to remove entire groups of nonsignificant terms at the same time. The statistical significance of individual terms can change because of other terms in the model. To reduce your model, you can use an automatic selection procedure, namely the *stepwise strategy*, to identify a useful subset of terms, choosing one of the three commonly used alternatives (standard stepwise, forward selection, and backward elimination).

Response Surface Regression: Mustard versus A; B; C; D

Analysis of Variance

Source	DF	Adj SS	Adj MS	F-Value	p-Value
Model	14	120.882	8.6344	5.55	0.000
Linear	4	76.960	19.2400	12.36	0.000
A	1	4.452	4.4524	2.86	0.096
B	1	34.836	34.8358	22.39	0.000
C	1	13.436	13.4362	8.63	0.005
D	1	21.910	21.9097	14.08	0.000
Square	4	10.654	2.6635	1.71	0.158
A*A	1	0.948	0.9476	0.61	0.438
B*B	1	8.873	8.8726	5.70	0.020
C*C	1	0.375	0.3755	0.24	0.625
D*D	1	0.788	0.7881	0.51	0.479
Two-way interaction	6	29.320	4.8866	3.14	0.009
A*B	1	0.506	0.5058	0.33	0.571
A*C	1	0.203	0.2035	0.13	0.719
A*D	1	0.039	0.0386	0.02	0.875
B*C	1	9.745	9.7450	6.26	0.015
B*D	1	6.758	6.7577	4.34	0.041
C*D	1	12.025	12.0251	7.73	0.007
Error	65	101.143	1.5560		
Lack-of-fit	25	46.991	1.8796	1.39	0.174
Pure error	40	54.152	1.3538		
Total	79	222.025			

3.5.2.3 Step 3 – If Need Be, Reduce the Model to Include the Significant Terms

To reduce the model, go to:

Stat > DOE > Response Surface > Analyze Response Surface Design

Select the numeric response variable "Mustard" under **Responses** and click on **Terms**. In **Include the following terms**, specify **Full Quadratic**, to study all main effects (both linear and quadratic effects) and the two-way interactions between factors. Note that in our example we don't have blocking factors (Stat Tool 2.4). If present, the grayed-out **Include blocks in the model option** would be selectable to consider blocks in the analysis. Click **OK** and in the main dialog box choose **Stepwise**. In **Method** select **Backward elimination** and in **Alpha to remove** specify **0.05**, then click **OK**. In the main dialog box choose **Graphs**. Under **Residual plots** choose **Four in one**. Minitab will display several residual plots to examine whether your model meets the assumptions of the analysis (Stat Tools 2.8, 2.9). Click **OK** in the main dialog box.

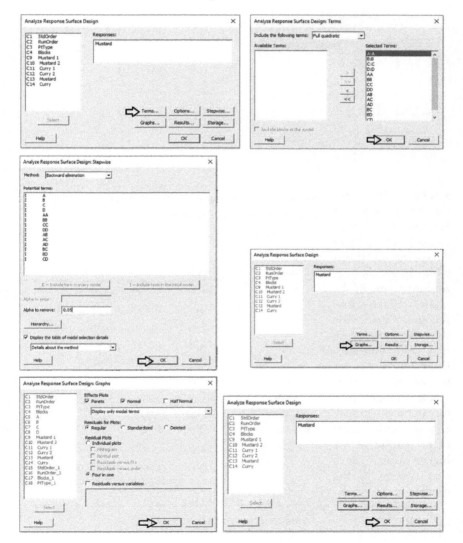

Complete the analysis by adding the factorial plots that show the relationships between the response and the significant terms in the model, displaying how the response mean changes as the factor levels or combinations of factor levels change.

To display factorial plots, go to:

Stat > DOE > Response Surface > Factorial Plots

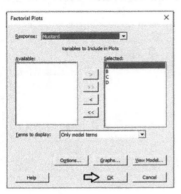

3.5.2.3.1 *Interpret the Results of Step 3*

Setting the significance level alpha to 0.05, the ANOVA table shows the significant terms in the model. Take into account that the stepwise procedure may add nonsignificant terms in order to create a hierarchical model. In a hierarchical model, all lower-order terms that comprise a higher-order term also appear in the model.

Note that in the ANOVA table, Minitab displays the

Response Surface Regression: Mustard versus A; B; C; D

Backward Elimination of Terms

α to remove = 0.05

Analysis of Variance

Source	DF	Adj SS	Adj MS	F-Value	P-Value
Model	6	107.690	17.948	11.46	0.000
Linear	3	80.045	26.682	17.04	0.000
B	1	38.869	38.869	24.82	0.000
C	1	14.717	14.717	9.40	0.003
D	1	25.803	25.803	16.47	0.000
Square	1	9.038	9.038	5.77	0.019
B*B	1	9.038	9.038	5.77	0.019
Two-way interaction	2	21.980	10.990	7.02	0.002
B*C	1	11.972	11.972	7.64	0.007
C*D	1	12.660	12.660	8.08	0.006
Error	73	114.336	1.566		
Lack-of-fit	33	60.184	1.824	1.35	0.183
Pure error	40	54.152	1.354		
Total	79	222.025			

lack-of-fit test when data contain replicates. This test helps investigators to determine whether the model accurately fits the data. Its null hypothesis states that the

Model Summary

S	R-sq	R-sq(adj)	R-sq(pred)
1.25150	48.50%	44.27%	37.37%

model accurately fits the data. As usual, compare the p-value to your significant level α. If the p-value is less than α, reject the null hypothesis and conclude that the lack of fit is statistically significant. If the p-value is larger than α, you fail to reject the null hypothesis and you can conclude that the lack of fit is not statistically significant.

In our example, the p-value (0.183) is greater than 0.05: we conclude that there is no strong evidence of lack of fit. Lack-of-fit can occur if important terms such as interactions or quadratic terms are not included in the model or if the fitted model shows several, unusually large residuals (Stat Tool 2.9).

In addition to the results of the analysis of variance, Minitab displays some other useful information in the Model Summary table.

The quantity **R-squared** (R-sq, R^2) is interpreted as the percentage of the variability among reflectance scores, explained by the terms included in the ANOVA model.

The value of R^2 varies from 0 to 100%, with larger values being more desirable.

The adjusted R^2 (R-sq(adj)) is a variation of the ordinary R^2 that is adjusted for the number of terms in the model. Use adjusted R^2 for more complex experiments with several factors, when you want to compare several models with different numbers of terms.

The value of **S** is a measure of the variability of the errors that we make when we use the ANOVA model to estimate the reflectance scores. Generally, the smaller it is, the better the fit of the model to the data.

Before proceeding with the results reported in the Session window, take a look at the residual plots. A **residual** represents an *error* that is the distance between an observed value of the response and its estimated value by the ANOVA model. The graphical analysis of residuals based on *residual plots* helps you to discover possible violations of the ANOVA underlying assumptions (Stat Tools 2.8, 2.9).

A check of the normality assumption could be made by looking at the histogram of the residuals (bottom left), but with small samples, often the histogram has an irregular shape.

The normal probability plot of the residuals may be more useful. Here we can see a tendency of the plot to bend slightly upward on the right, but the plot is not in any case grossly non-normal. In general, moderate departures from normality are of little concern in the ANOVA model with fixed effects.

The other two graphs (Residuals versus Fits and Residuals versus Order) seem unstructured, thus supporting the validity of the ANOVA assumptions.

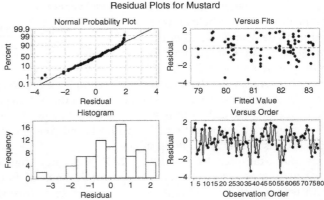

Residual Plots for Mustard

Returning to the Session window, we find a table of coded coefficients followed by a regression equation in uncoded units.

In ANOVA, we have seen that many of the questions the experimenter wishes to answer can be solved by several hypothesis tests.

It is also helpful to present the results in terms of a *regression model*, i.e. an equation derived from the data that expresses the relationship between the response and the important design factors.

In the regression equation, we have a coefficient for each term

Coded Coefficients

Term	Coef	SE Coef	t-Value	p-Value	VIF
Constant	81.801	0.249	328.05	0.000	
B	0.855	0.172	4.98	0.000	1.01
C	0.543	0.177	3.07	0.003	1.02
D	0.695	0.171	4.06	0.000	1.04
B*B	−0.726	0.302	−2.40	0.019	1.02
B*C	0.600	0.217	2.76	0.007	1.02
C*D	−0.610	0.215	−2.84	0.006	1.05

Regression Equation in Uncoded Units

$$Mustard = 78.971 + 11.38B + 0.369C$$
$$+0.2611D - 32.2B*B + 2.665B*C$$
$$-0.0814C*D$$

in the model. They are expressed in *uncoded units,* i.e. in the original measurement units. These coefficients are also displayed in the Coded Coefficients table, but in *coded units.* Earlier, we learned how to examine the p-values in the ANOVA table to determine whether a factor is statistically related to the response. In the Coded Coefficients table, we find the same results in terms of tests on coefficients and we can evaluate the p-values to determine whether they are statistically significant. The null hypothesis states the absence of any effect on the response. Setting the significance level α at 0.05, if a p-value is less than α, reject the null hypothesis and conclude that the effect is statistically significant.

Look at the factorial plots to better understand how the mean reflectance score varies as factor levels change.

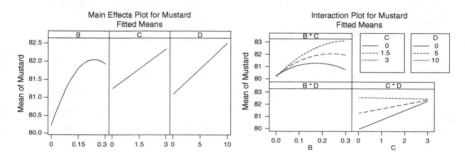

The presence of the quadratic term B*B makes a curvature in the factorial plots involving component B. Considering the interaction B*C, the best result (the higher mean cleaning score) is obtained when B and C are increasing in value. For the interaction C*D, when C is equal to its high*est* level (3), the mean cleaning score does not vary much across levels of component D. When C is absent, the mean cleaning score tends to increase with increasing levels of D. Note that the interaction C*D lies on the left side of the **normal plot of the standardized effects**.

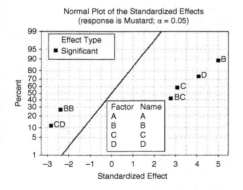

Now repeat steps 2 and 3 for the second response variable "Curry." For brevity, we show only the reduced model with residual and factorial plots.

Response Surface Regression: Curry versus A; B; C; D

Backward Elimination of Terms

α to remove = 0.05

Analysis of Variance

Source	DF	Adj SS	Adj MS	F-Value	p-Value
Model	7	174.552	24.9361	26.20	0.000
Linear	4	134.193	33.5482	35.24	0.000
A	1	5.886	5.8857	6.18	0.015
B	1	57.281	57.2815	60.18	0.000
C	1	7.154	7.1544	7.52	0.008
D	1	59.144	59.1438	62.13	0.000
Square	1	5.807	5.8071	6.10	0.016
B*B	1	5.807	5.8071	6.10	0.016
Two-way interaction	2	45.011	22.5055	23.64	0.000
B*C	1	18.653	18.6527	19.60	0.000
C*D	1	31.251	31.2506	32.83	0.000
Error	72	68.536	0.9519		
Lack-of-fit	32	29.970	0.9366	0.97	0.529
Pure error	40	38.566	0.9642		
Total	79	243.089			

Model Summary

S	R-sq	R-sq(adj)	R-sq(pred)
0.975651	71.81%	69.06%	65.43%

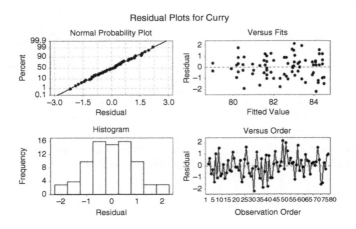

Residual Plots for Curry

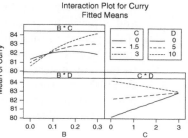

3.5.2.4 Step 4 – Optimize the Responses

After fitting the response surface models, additional analyses can be carried out to find the optimum set of formulation components that optimizes the response variables (Stat Tool 3.8).

To optimize the response variables, go to:

Stat > DOE > Response Surface > Response Optimizer

In the table, under **Goal**, select one of the following options for each response:

- **Do not optimize**: Do not include the response in the optimization process.
- **Minimize**: Lower values of the response are preferable.
- **Target**: The response is optimal when values meet a specific target value.
- **Maximize**: Higher values of the response are preferable.

In our case study, the goal is to maximize the reflectance data: choose **Maximize** from the drop-down list and click **Setup**. In the next window you need to specify a response target and lower or upper bounds, depending on your goal. Minitab automatically sets the **Lower value** and the **Upper value** to the smallest and largest values in the data, but you can change these values depending on your response goal (Stat Tool 3.8).

- If your goal is to maximize the response (larger is better), you need to specify the lower bound and a target value. You may want to set the lower value to the smallest acceptable value and the target value at the *point of diminishing returns*, i.e. a point at which going above its value does not make much difference. If there is no point of diminishing returns, use a very high target value.

- If your goal is to target a specific value, you should set the lower value and upper value as points of diminishing returns.
- If the goal is to minimize (smaller is best), you need to specify the upper bound and a target value. You may want to set the upper value to the largest acceptable value and the target value at the point of diminishing returns, i.e. a point at which going below its value does not make much difference. If there is no point of diminishing returns, use a very small target value.

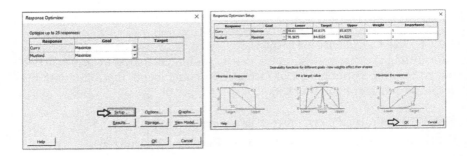

In our example, we leave the default values with the lower bound set to the minimum for each response and the target and upper values set to the maximum.

Note that in the process of response optimization, Minitab calculates and uses desirability indicators from 0 to 1 to represent how well a combination of factor levels satisfies the goals you have for the responses. Specifying a different *weight* varying from 0.1 to 10, you can emphasize or deemphasize the importance of attaining the goals. Furthermore, when you have more than one response, Minitab calculates individual desirabilities for each response and a composite desirability for the entire set of responses. With multiple responses, you may assign a different level of *importance* to each response by specifying how much effect each response has on the composite desirability.

In our example, leave **Weight** and **Importance** equal to **1** and click **OK**. In the main dialog box, select **Results** and in **Number of solutions to display**, specify **5**. Click **OK** in each dialog box. In the optimization plot, Minitab will display the optimal solution, that is the factor setting which maximizes the composite desirability, while in the Session window we will find a list of five different factor combinations close to the optimal one, which is the first on the list.

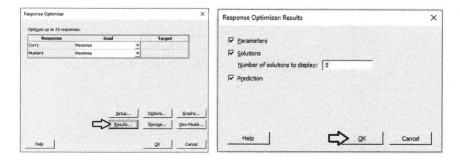

The optimization plot shows how different factor settings affect the responses. The vertical red lines on the graph represent the current factor settings. The numbers displayed at the top of a column show the current factor level settings (in red). The horizontal blue lines and numbers represent the responses for the current factor level. The optimal solution is plot on the graph and serves as the initial point. The settings can then be modified interactively, to determine how different combinations of factor levels affect responses, by moving the red vertical lines corresponding to each factor. To return to the initial settings, right-click and select **Navigation** and then **Reset to Initial Settings**.

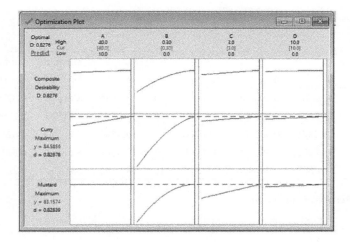

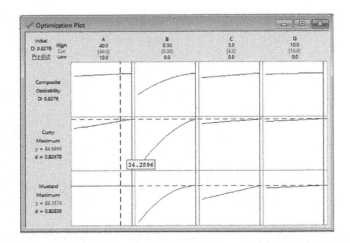

3.5.2.4.1 Interpret the Results of Step 4 The optimization plot is a useful tool to explore the sensitivity of response variables to changes in the factor settings and in the vicinity of a local solution – for example, the optimal one.

Response Optimization: Curry; Mustard

Parameters

Response	Goal	Lower	Target	Upper	Weight	Importance
Curry	Maximum	78.6100	85.8375		1	1
Mustard	Maximum	76.5675	84.5225		1	1

Solutions

					Curry	Mustard	Composite
Solution	A	B	C	D	Fit	Fit	Desirability
1	40.0000	0.300000	3.00000	10	84.5856	83.1574	0.827588
2	39.5859	0.300000	3.00000	10	84.5763	83.1574	0.826945
3	39.2871	0.300000	3.00000	10	84.5696	83.1574	0.826481
4	39.2056	0.300000	3.00000	10	84.5678	83.1574	0.826354
5	40.0000	0.240593	2.80211	10	84.2634	83.0032	0.795500

Multiple Response Prediction

Variable	Setting
A	40
B	0.3
C	3
D	10

Response	Fit	SE Fit	95% CI	95% PI
Curry	84.586	0.344	(83.901; 85.271)	(82.524; 86.648)
Mustard	83.157	0.419	(82.322; 83.993)	(80.527; 85.788)

The Session window displays:

- Information about the boundaries, weight, and importance for each response variable (Parameters).
- The factor settings, the fitted responses, and the composite desirability for the required 5 solutions. The first in the list is the optimal solution (Solutions).
- The factor settings for the optimal solution and the fitted responses with their confidence intervals (Stat Tool 1.14) and prediction intervals (Multiple Response Prediction).

For reflectance data, the optimal solution is the formulation with component A set to 40, B to 0.30, C to 3, and D to 10. For this setting, the fitted response is equal to 84.59 (95% CI: 83.90–85.27) for the Curry stain and 83.16 (95% CI: 82.32–83.99) for Mustard. The composite desirability (0.83) is reasonably close to 1, denoting a satisfactory overall achievement of the specified goals.

Stat Tool 3.8 Response Optimization

In a mature phase of an experimental work, an important objective is the *optimization* of the responses, i.e. the search of those combinations of factor levels that are related to the desirable value of response variables. When at least one factor is a quantitative variable, *response surface methodology* is useful to study in depth how several factors affect one or more response variables and how to optimize these responses. After building an appropriate *response surface model* (generally a *full quadratic model* including linear and quadratic effects of factors and the two-way interactions) that expresses the relationship between a response and the factors, a *response optimization approach* based on the use of *desirability functions* helps identify the combination of factor settings that optimizes a single response or a set of responses (Figure 3.1).

For a response variable to optimize, a desirability function varying from 0 to 1 is calculated to represent how well a combination of factor levels satisfies a specific goal set for the response. This goal may be defined in the following way:

- Minimize the response (smaller values of the response are better).
- Target the response (a specific target value for the response is desirable).
- Maximize the response (higher values are better).

A desirability equal to 1 represents the ideal case (e.g. hitting a target value for the response). A desirability of zero indicates that the response is outside of the acceptable values. Any values within the acceptable values will result in a desirability greater than zero.

Stat Tool 3.8 (Continued)

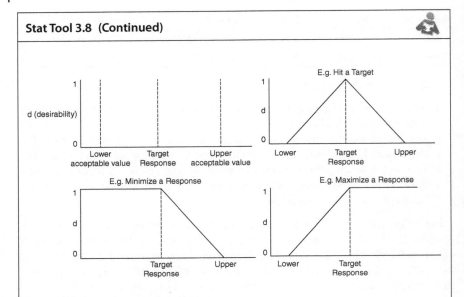

Figure 3.1 Desirability plots with different goals.

If there are multiple responses, you can specify a different goal for each response. Individual desirabilities related to single responses are then combined to determine the *composite desirability* for the entire set of responses. An *optimal solution* (that is an optimal factor setting) occurs where composite desirability obtains its maximum.

3.5.2.5 Step 5 – Examine the Shape of the Response Surface and Locate the Optimum

Once investigators have found the optimal solutions, it is usually convenient to study the response surface in the vicinity of these points. Graphic techniques such as *contour*, *surface*, and *overlaid plots* help you to examine the shape of the response surface in regions of interest. Through these plots, you can explore the potential relationship between *pairs of factors* and the responses. You can display a contour, a surface, or an overlaid plot considering two factors at a time, while setting the remaining factors to predefined levels. With contour and surface plots, you consider a single response; in overlaid plots, you consider all the responses together.

To display the contour plots, go to: **Stat > DOE > Response Surface > Contour Plot**

In the dialog box, select the first response "Mustard" in **Response**, and check the option **Generate plots for all pairs of continuous variables**. Click **Contours**, and under **Data Display**, choose **Contour lines**. Click **OK** and in the main dialog box and select **Settings**. Set the factors at levels shown in the first optimal solution found previously in the Response Optimizer (A = 40, B = 0.30, C = 3, D = 10). Remember that for the response Mustard, only three components B, C, D showed a significant relationship with this response. For this reason, in the window **Contour Plot: Settings**, you must specify values only for these three factors. Click **OK** in each dialog box. Do the same for the response variable "Curry."

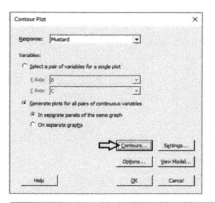

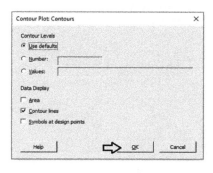

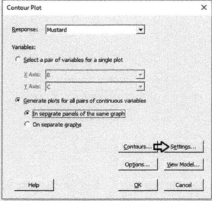

To display the surface plots, go to: **Stat > DOE > Response Surface > Surface Plot**

In the dialog box, select the first response "Mustard" in **Response**, and check the option **Generate plots for all pairs of continuous variables**. Then click **Settings**, and set the factors at levels shown in the first optimal solution found previously in the Response Optimizer (A = 40, B = 0.30, C = 3, D = 10). Click **OK** in each dialog box. Do the same for the response variable "Curry."

To change the appearance of the surface plots, double-click on any point inside a surface. In **Attributes**, under **Surface Type**, select the option **Wireframe** instead of the default **Surface**.

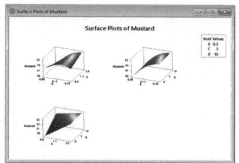
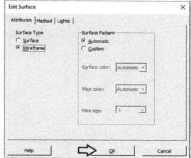

To display the overlaid plots, go to: **Stat > DOE > Response Surface > Overlaid Contour Plot**

Select all the responses and click **Contours**. Note that we are building the overlaid contour plot for the pair A and B. In the next window, for each response, you need to define a low and a high contour, depending on your goal for the responses. For example, you may want to set the **Low** value and the **High** value at the points of diminishing returns, or you can simply consider the fitted response values obtained with the optimal solution and specify two points close to the fitted values. In step 4, the fitted response values for the optimal solution were 83.157 for Mustard and 84.586 for Curry. Specify the interval **82–84** for Curry and **81–83** for Mustard as **Contours** and then click **OK**. In the main dialog box, select **Settings**. Set the remaining factors (C and D) at levels shown in the first optimal solution found previously in the Response Optimizer (A = 40, B = 0.30, C = 3, D = 10). Click **OK** in each dialog box. Repeat the process for the other factor pairs to create all overlaid contour plots.

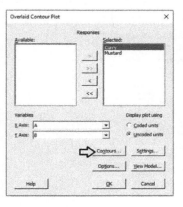
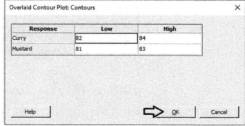

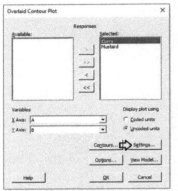 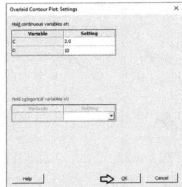

3.5.2.5.1 *Interpret the Results of Step 5* Contour and surface plots are useful tools to explore how a response variable changes while increasing or decreasing in factor levels. In contour plots, two factors are plotted on the *x*- and *y*- axes and the reflectance performance is represented by different level curves, while the remaining factors are set to predefined levels (Hold values). Double-click on any point in the frame and select **Crosshairs** to interactively view variations in the response. In surface plots the reflectance performance is represented by a smooth surface.

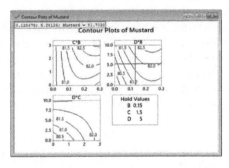

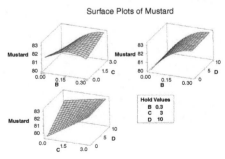

Surface Plots of Mustard

Contour Plots of Curry

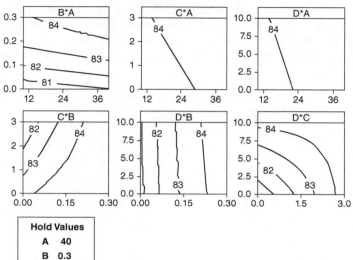

Hold Values
A 40
B 0.3
C 3
D 10

Surface Plots of Curry

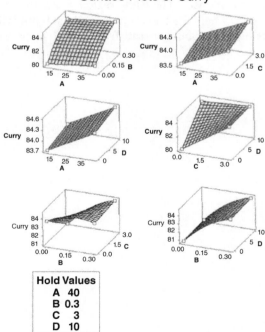

Hold Values
A 40
B 0.3
C 3
D 10

In overlaid contour plots we simultaneously consider both responses in order to explore factor settings that to a certain extent optimize all responses or keep them in desired ranges of values. Remember that Minitab builds an overlaid contour plot for each factor pair. For brevity, here we show only one of these plots, namely that related to factors C and D, respectively plotted on the x- and y-axes.

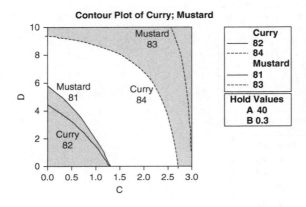

Double-click on any point in the frame and select **Crosshairs** to interactively view variations in the responses. The white region shows the ranges of values for component C and D where the criteria (low and high boundaries previously defined in the **Settings** dialog box) are satisfied for both response variables.

4

Other Topics in Product Development and Optimization: Response Surface and Mixture Designs

4.1 Introduction

In the previous chapter, the case study on the development of an innovative dishwashing product discussed the design and analysis of a general factorial design to identify which formula maximizes the cleaning performance response.

The present chapter extends the topic of design and analysis of factorial designs by introducing two classes of designs particularly suited to fitting and analyzing response surfaces and for optimizing response variables: central composite designs (CCDs) and Box-Behnken designs.

When the factors represent ingredients or components of a formulation, the total amount of which is fixed and we are interested in evaluating whether one or more responses vary when decreasing or increasing the proportions of the ingredients, mixture experiments can be designed and analyzed to take into account the dependence among components.

The first case study considers the development of a new polymeric membrane, with the goal being to maximize the product's elasticity. Based on previous results, the scientists identify the region of exploration to search for the factor levels that maximize product elasticity. The example shows how to plan a CCD or a Box-Behnken design in line with specific interests. For the analysis of the design, given that the experimenters are relatively close to the optimum set of factor levels, a second-order model is fitted to allow the detection of curvatures, if any, in the response surface.

The second example considers a mixture experiment. In the development of a new detergent, the investigators aim to analyze to what extent three different components of a formulation can affect the product's technical performance. The reader is guided through the planning of different kinds of mixture designs depending on the presence or absence of other process variables in the design,

End-to-End Data Analytics for Product Development: A Practical Guide for Fast Consumer Goods Companies, Chemical Industry and Processing Tools Manufacturers, First Edition.
Rosa Arboretti, Mattia De Dominicis, Chris Jones, and Luigi Salmaso
© 2020 John Wiley & Sons Ltd. Published 2020 by John Wiley & Sons Ltd.
Companion website: www.wiley.com/go/salmaso/data-analytics-for-pd

or the presence of component bounds or linear constraints among components. The analysis of the selected mixture design closes the chapter.

In short, the chapter deals with the following:

Topics	Stat tools
Central composite designs	4.1
Box-Behnken designs	4.2
Mixture experiments	4.3
Simplex centroid designs	4.4
Simplex lattice designs	4.5
Constrained simplex designs	4.6
Extreme vertices designs	4.7
Mixture models	4.8

Learning Objectives and Outcomes

Upon completion of this chapter, you should be able to do the following:

Know how to plan designs to fit second-order models.
Understand the differences between central composite designs and Box-Behnken designs.
Be able to recognize mixture experiments.
Understand the differences between different kinds of mixture designs.
Know how to plan and analyze mixture experiments.

4.2 Case Study for Response Surface Designs: Polymer Project

In the development of a new polymeric membrane, the research team aims to maximize the product's elasticity. Previous research has shown that the concentrations of two additives (factors A, B) and the exposure temperature explain much of the variability in elasticity. The scientists decide that the region of exploration in the search for factor levels that maximize the elasticity of the product should be around (3%, 10%) for the two additives and (40°, 80°) for the temperature. For logistic reasons, investigators need to arrange the experiment in such a way that the combinations of factor levels will be tested by two different operators. As the experimenters are relatively close to having the optimum set of factor levels, a second-order model will be suitable to study all main effects (both linear and quadratic) and the two-way interactions between factors. This model will allow the detection of curvatures in the response surface if any.

4.2.1 Plan of the Experimental Design

In Chapter 3, section 3.5, we fit a second-order model to analyze a general factorial design with continuous factors with more than two levels *specified by the experimenters* and estimate the response surface in order to identify regions where the responses were close to the optimum. Now, as an alternative approach, we will create the experimental design by using a popular class of response surface designs called *CCD* (Stat Tool 4.1). Then we will present a variant of the CCD called the *face-centered design,* and lastly another standard response surface design named the *Box-Behnken design* (Stat Tool 4.2). The experimenters will choose one of the three designs according to the characteristics of each one and the main purposes of the study.

For this study:

- The *two additives and the temperature* represent the *key factors.*
- For factors A and B, the *percentages of concentrations* 3% and 10% represent the *factor levels* for the definition of the *cube points.*
- For factor Temperature, the temperatures 40° and 80° represent the *factor levels* for the definition of the *cube points.*
- The *elasticity* represents *the response variable.*
- The *two operators* represent *two blocks* (Stat Tool 2.4).

To perform the experiment, let's proceed in the following way:

Step 1 – Create a CCD.
Step 2 – *Alt*ernatively, create a face-centered CCD.
Step 3 – *Alt*ernatively, create a Box-Behnken design.
Step 4 – Assign the designed factor-level combinations (design points) to the experimental units and collect data for the response variable.

4.2.1.1 Step 1 – Create a CCD

Let's begin creating a CCD, considering the three factors A, B, and Temperature.

To create a CCD, go to: **Stat > DOE > Response Surface > Create Response Surface Design** Under **Type of Design** select **Central composite** and in **Number of continuous factors** specify 3. Note that you can also consider categorical factors in your design.

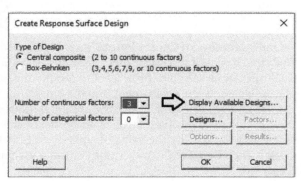

Click on **Display Available Designs** to have an idea of the number of runs based on the design type both with and without blocks and the number of continuous factors in the design. Note that depending on the number of continuous factors, you can create a full or fractional design (Stat Tool 2.2). Click on **Designs**. In the next dialog box, select the number of experimental runs and the number of blocks for the design. Note that with two blocks (the two operators), the CCD will include 20 runs.

There will be six center points, four of which will be included in the cube block, i.e. in the block with the experimental runs at the high and low levels of the factors. The other two center points will be in the axial block, i.e. the other block with the experimental runs at the axial levels of the factors.

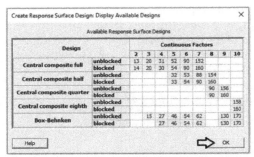

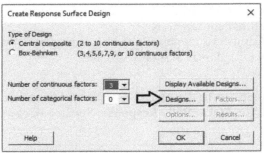

Under **Number of Center Points** leave the selected option **Default** to ensure that the design preserves some desirable statistical properties. However, **Custom** can be selected if you want to modify the number of center points in the design and their deployment into the blocks.

Under **Value of Alpha** check the option **Default** to ensure that the design preserves some desirable statistical properties. Again, **Custom** can be selected if you want to modify the value of alpha used to define the axial points. The option **Face Centered** allows you to create a variant of the CCD called Face Centered CCD (Stat Tool 4.1).

In **Number of replicates** specify **1** or, depending on available resources, enter how many times to perform each experimental run. Remember you can add replicates to your design later with **Stat > DOE > Modify Design.** If you specify more than 1 replicate and check the option **Block on replicates,** Minitab will put each replicate into its own block. If the base design has blocks, then the total number of blocks will be the product of the number of blocks in the base design and the number of replicates. Click **OK.**

In the main dialog box select **Factors.** Under **Name, Low** and **High,** enter a name for each factor and its high and low levels. By default, Minitab sets the low level of all factors to –1 and the high level to +1. Under **Levels Define,** choose the appropriate factor setting **Cube points** or **Axial points.** In our example, the factor setting represents Cube points. If you have categorical factors, the window will allow you to enter the name, number of levels, and level values.

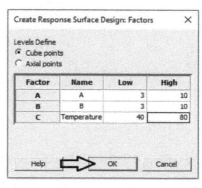

Proceed by clicking **Options** in the main dialog box and check the option **Randomize runs** (Stat Tool 2.3). Click **OK** in the main dialog box and Minitab shows the design of the experiment in the worksheet, and some information in the Session window.

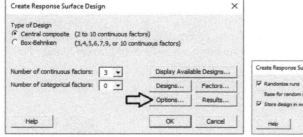
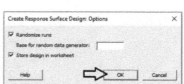

You can see that the design includes 20 runs distributed between two blocks. Column PtType shows each design point's type, where the value "0" represents a central point, "1" a cube point, and "−1" an axial point. Block 1 (one of the two operators) is a cube block since it includes cube points with the factors set at their low or high levels. Block 2 is an axial block since it includes axial points. Four center points are in block 1 and two in block 2.

↓	C1	C2	C3	C4	C5	C6	C7	C8
	StdOrder	RunOrder	PtType	Blocks	A	B	Temperature	
1	20	1	0	2	6.5000	6.5000	60.00	
2	18	2	-1	2	6.5000	6.5000	92.66	
3	13	3	-1	2	0.7845	6.5000	60.00	
4	14	4	-1	2	12.2155	6.5000	60.00	
5	16	5	-1	2	6.5000	12.2155	60.00	
6	19	6	0	2	6.5000	6.5000	60.00	
7	15	7	-1	2	6.5000	0.7845	60.00	
8	17	8	-1	2	6.5000	6.5000	27.34	
9	9	9	0	1	6.5000	6.5000	60.00	
10	10	10	0	1	6.5000	6.5000	60.00	
11	8	11	1	1	10.0000	10.0000	80.00	
12	11	12	0	1	6.5000	6.5000	60.00	
13	1	13	1	1	3.0000	3.0000	40.00	
14	7	14	1	1	3.0000	10.0000	80.00	
15	3	15	1	1	3.0000	10.0000	40.00	
16	4	16	1	1	10.0000	10.0000	40.00	
17	5	17	1	1	3.0000	3.0000	80.00	
18	6	18	1	1	10.0000	3.0000	80.00	
19	2	19	1	1	10.0000	3.0000	40.00	
20	12	20	0	1	6.5000	6.5000	60.00	
21								

Central Composite Design

Design Summary

Factors:	3	Replicates:	1
Base runs:	20	Total runs:	20
Base blocks:	2	Total blocks:	2

$\alpha = 1.633$

Two-level factorial: Full factorial

Point Types

Cube points:	8
Center points in cube:	4
Axial points:	6
Center points in axial:	2

Stat Tool 4.1 The Central Composite Design (CCD)

The search for a suitable approximating function for the relationship between one or more responses and a set of factors (some being quantitative in nature) is useful to study in depth how factors affect response variables and how to optimize these responses. A response surface represents the functional relationship between responses and factors. Generally, in the mature phase of an experiment, second-order models, including both linear and quadratic effects and the two-way interactions between factors, allow the detection of curvatures, if any, in the response surface.

This response surface is represented in Figure 4.1 without curvatures (graph A), when the functional relationship is approximated by a first-order model, and with curvatures (graph B), when the functional relationship is approximated by a second-order model.

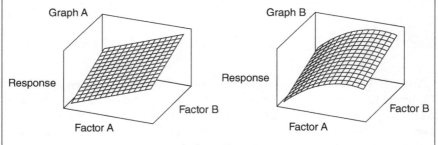

Figure 4.1 Response surface plot for a first-order (linear) model (graph A) and for a second-order model with curvatures (graph B).

Stat Tool 4.1 (Continued)

A popular class of designs used to fit second-order models are the CCDs. A CCD is usually used after having performed a factorial or fractional factorial experiment and having identified the key factors. It requires the experimenter to identify *only two levels* (a low and a high level) for each factor. These levels are not the minimum and maximum admissible values for the factors, but are values *around which* the experimenter wants to explore the response surface. Generally, the CCD consists of a 2^k factorial design (based on the two levels defined by the experimenter for each factor, with k equal to the number of factors) with the addition of:

- a certain number of replicates of the *center point* (where all the factors are set at their middle value between the low and high levels);
- $2k$ *axial* or *star points* (where all the factors are set at their middle point except one that is placed at a specific distance "α" from its middle value).

CCDs are especially useful in *sequential experiments* because you can often build on previous full factorial or fractional factorial designs by adding center points and axial points.

Figure 4.2 shows the CCD for $k = 2$ and $k = 3$ with the low and high levels set at "−1" and "+1."

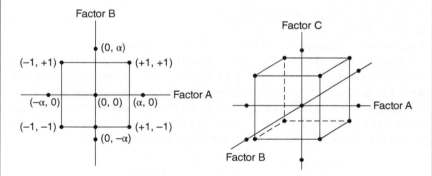

Figure 4.2 Central composite designs for $k = 2$ and $k = 3$.

For example, with two factors A and B, the CCD includes the following:

- The factorial design points (−1,−1), (+1,−1), (−1,+1), (+1,+1) represent the vertices of the so-called "cube" around which the design is built.
- Some replicates of the center point (0, 0), with the two factors set at their middle value between the low and high levels.

Stat Tool 4.1 (Continued)

- Four axial points $(-\alpha, 0)$, $(\alpha, 0)$, $(0, -\alpha)$, $(0, \alpha)$, where one factor is set at its middle value and the other factor is set at a distance "α" from its middle value. If "α" is less than 1, the axial points are inside the "cube"; if "α" is greater than 1, the axial points are outside the "cube"; if "α" is equal to 1, the axial points are on the face of the "cube."

Note that the axial points are usually outside the "cube" (unless you specify an α that is less than or equal to 1). If you are not careful, this could lead to axial points that are not in the region of interest or may be impossible to test.

In many situations, the region of interest in which to explore the relationship between factors and responses, may be imagined as a *sphere* around the basic factorial design. In these cases, the best choice for the distance "α" for the definition of the axial points is $\alpha = \sqrt{k}$, with k equal to the number of factors. This design includes factorial and axial design points lying on the surface of a sphere of radius $\sqrt{k}$ (Figure 4.3).

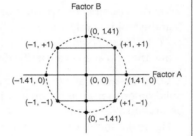

Figure 4.3 Spherical CCD for $k = 2$ with $\alpha = \sqrt{2}$.

For a *spherical CCD*, three to five runs of the *center point* where all the factors are set at their middle value (for example with $k = 2$, the point denoted by "0,0"), are generally recommended.

When the region of interest is *cuboidal* rather than spherical, a useful variation of the central composite design is the *face-centered central composite design*, in which $\alpha = 1$. In this design the axial points are located on the centers of the faces of the cube (Figure 4.4). For a face-centered CCD, two or three runs of the central point are generally sufficient. The face-centered CCD requires only three levels of each factor. For this reason it is often used when factor levels are difficult to change.

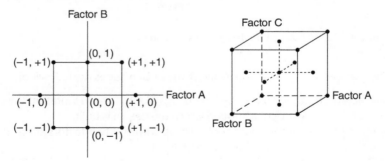

Figure 4.4 Face-centered central composite designs for $k = 2$ and $k = 3$.

4.2.1.2 Step 2 – Alternatively, Create a Face-Centered CCD

If you need to evaluate the relationship between the three factors (A, B, Temperature) and elasticity, by exploring a cuboidal region around the basic factorial design, you can create a face-centered CCD (Stat Tool 4.1).

 To create a face-centered CCD, go to: **Stat > DOE > Response Surface > Create Response Surface Design**

Under **Type of Design** select **Central composite** and in **Number of continuous factors** specify 3.

Click on **Designs**. In the next dialog box, select the number of experimental runs and the number of blocks for the design. As outlined in step 1, with two blocks (the two operators), the CCD will include 20 runs. There will be six center points, four of which will be included in the cube block, i.e. in the block with the experimental runs at the high and low levels of the factors. The other two center points will be in the axial block, i.e. the other block with the experimental runs at the axial levels of the factors.

Under **Number of Center Points** leave the selected option **Default** to ensure that the design preserves some desirable statistical properties. However, **Custom** can be selected if you want to modify the number of center points in the design and their deployment into the blocks.

Under **Value of Alpha** check the option **Face Centered**.

In **Number of replicates** specify 1 or, depending on available resources, enter how many times to perform each experimental run. Remember you can add replicates to your design later with **Stat > DOE > Modify Design**. If you specify more than 1 replicate and check the option **Block on replicates**, Minitab will put each replicate into its own block. If the base design has blocks, then the total number of blocks will be the product of the number of blocks in the base design and the number of replicates. Click **OK**.

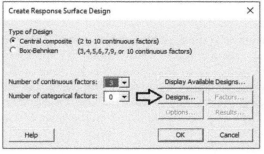

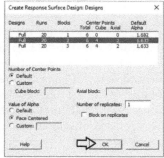

In the main dialog box select **Factors**. Under **Name**, **Low** and **High**, enter a name for each factor and its high and low levels. By default, Minitab sets the low level of all factors to −1 and the high level to +1. Under **Levels Define**, choose the appropriate factor setting **Cube points** or **Axial points**. In our example, the factor setting represents Cube points. If you have categorical factors, the window will allow you to enter the name, number of levels, and level values.

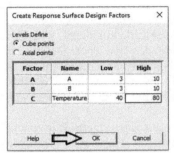
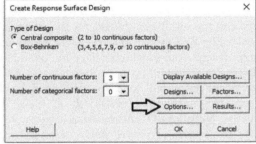

Proceed by clicking **Options** in the main dialog box and check the option **Randomize runs** (Stat Tool 2.3). Click **OK** in the main dialog box and Minitab shows the design of the experiment in the worksheet, and some information in the Session window.

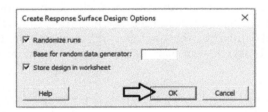

You can see that the design includes 20 runs distributed between two blocks. Column PtType shows each design point's type, where the value "0" represents a central point, "1" a cube point, and "−1" an axial point. Block 1 (one of the two operators) is a cube block since it includes cube points with the factors set at their low or high levels. Block 2 is an axial block since it includes axial points. Four center points are in block 1 and two in block 2. Note that the value of α is now set to 1.

▲	C1 StdOrder	C2 RunOrder	C3 PtType	C4 Blocks	C5 A	C6 B	C7 Temperature	C8
1	5	1	1	1	3.0	3.0	80	
2	6	2	1	1	10.0	3.0	80	
3	8	3	1	1	10.0	10.0	80	
4	4	4	1	1	10.0	10.0	40	
5	9	5	0	1	6.5	6.5	60	
6	2	6	1	1	10.0	3.0	40	
7	3	7	1	1	3.0	10.0	40	
8	7	8	1	1	3.0	10.0	80	
9	1	9	1	1	3.0	3.0	40	
10	12	10	0	1	6.5	6.5	60	
11	10	11	0	1	6.5	6.5	60	
12	11	12	0	1	6.5	6.5	60	
13	17	13	-1	2	6.5	6.5	40	
14	20	14	0	2	6.5	6.5	60	
15	16	15	-1	2	6.5	10.0	60	
16	14	16	-1	2	10.0	6.5	60	
17	18	17	-1	2	6.5	6.5	80	
18	19	18	0	2	6.5	6.5	60	
19	15	19	-1	2	6.5	3.0	60	
20	13	20	-1	2	3.0	6.5	60	
21								

Central Composite Design

Design Summary

Factors:	3	Replicates:	1
Base runs:	20	Total runs:	20
Base blocks:	2	Total blocks:	2

$\alpha = 1$

Two-level factorial: Full factorial

Point Types

Cube points:	8
Center points in cube:	4
Axial points:	6
Center points in axial:	2

4.2.1.3 Step 3 – Alternatively, Create a Box-Behnken Design

You can decide to evaluate the relationship between the three factors (A, B, Temperature) and elasticity, by creating a Box-Behnken design (Stat Tool 4.2), which often has fewer design points than CCDs with the same number of factors. The number of blocks (Stat Tool 2.4) that it is possible to include in a Box-Behnken design depends on the number of factors. Designs with three factors cannot have blocks. For this reason, we now suppose the absence of blocks in our experiment.

 To create a Box-Behnken design, go to: **Stat > DOE > Response Surface > Create Response Surface Design**

Under **Type of Design** select **Box-Behnken** and in **Number of continuous factors** specify 3.

Click on **Designs**. Under **Number of Center Points** you can leave the selected option **Default**. However, select **Custom** if you wish to modify the number of center points in the design. In Number of replicates specify **1** or, depending on available resources, enter how many times to perform each experimental run. Remember you can add replicates to your design later with **Stat > DOE > Modify Design**. If you specify more than 1 replicate, check the option **Block on replicates** and Minitab will put each replicate into its own block. If the base design has blocks, then the total number of blocks will be the product of the number of blocks in the base design and the number of replicates. If you have blocks (with more than three factors), then specify the number of blocks and click **OK**.

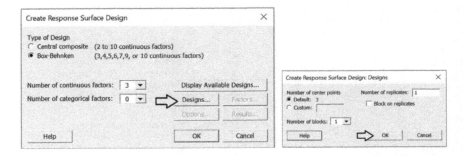

In the main dialog box select **Factors**. Under **Name, Low** and **High**, enter a name for each factor and its high and low levels. By default, Minitab sets the low level of all factors to −1 and the high level to +1. If you have categorical factors, the window will allow you to enter the name, number of levels, and level values.

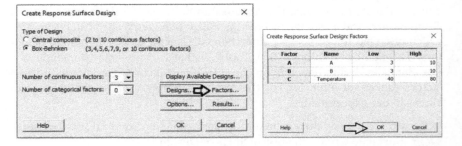

Proceed by clicking **Options** in the main dialog box and select **Randomize runs** (Stat Tool 2.3). Click **OK** in the main dialog box and Minitab shows the design of the experiment in the worksheet, as well as some information in the Session window.

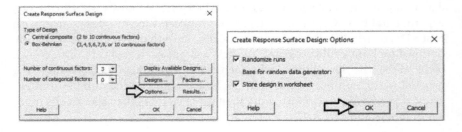

You can see the design includes 15 runs. Column PtType shows each design point type, where the value "0" represents a central point and "2" an edge mid-point.

.	C1	C2	C3	C4	C5	C6	C7	C8
	StdOrder	RunOrder	PtType	Blocks	A	B	Temperature	
1	6	1	2	1	10,0	6,5	40	
2	5	2	2	1	3,0	6,5	40	
3	14	3	0	1	6,5	6,5	60	
4	15	4	0	1	6,5	6,5	60	
5	13	5	0	1	6,5	6,5	60	
6	8	6	2	1	10,0	6,5	80	
7	4	7	2	1	10,0	10,0	60	
8	11	8	2	1	6,5	3,0	80	
9	9	9	2	1	6,5	3,0	40	
10	12	10	2	1	6,5	10,0	80	
11	1	11	2	1	3,0	3,0	60	
12	3	12	2	1	3,0	10,0	60	
13	2	13	2	1	10,0	3,0	60	
14	7	14	2	1	3,0	6,5	80	
15	10	15	2	1	6,5	10,0	40	
16								

Box-Behnken Design

Design Summary

Factors:	3	Replicates:	1
Base runs:	15	Total runs:	15
Base blocks:	1	Total blocks:	1

Center points: 3

Stat Tool 4.2 The Box-Behnken Design

The Box-Behnken design is a three-level design with no axial points and no cube points. It considers combinations of factor levels where factors are set at their low or high levels and one factor is at its middle value (edge midpoints), and some replicates of the center point. Figure 4.5 shows the Box-Behnken design with three factors, denoting the low and high values as "−1" and "+1." Notice that the design includes the edge midpoints lying on a sphere of radius $\sqrt{2}$. It often has fewer design points than CCDs with the same number of factors.

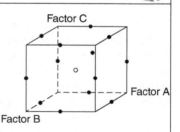

Figure 4.5 The Box-Behnken design for $k = 3$.

The following table presents a comparison of CCD and Box-Behnken designs, along with some useful notes for their implementation in Minitab.

Central composite designs	Box-Behnken designs
For k factors, they consist of: - 2^k full factorial (or fractional factorial) - $2k$ axial or star design points (axial runs) - a number of center design points They have axial points, which therefore may not be in the region of interest or may be impossible to test.	For k factors, they consist of: - edge midpoints - a number of center design points They do not have axial points; thus, all design points fall within your safe operating zone.

Stat Tool 4.2 (Continued)

Central composite designs	Box-Behnken designs
They include runs where all factors are at their extreme setting, such as all at low levels.	They never include runs where all factors are at their extreme setting, such as all at low levels.
In Minitab you can specify 2–10 continuous factors.	In Minitab you can specify 3–10 continuous factors.
In Minitab, starting with the specification of two-level factors, the design can include up to five levels per factor.	In Minitab, starting with the specification of two-level factors, the design can include up to three levels per factor.

4.2.1.4   Step 4 – Assign the Designed Factor Level Combinations to the Experimental Units and Collect Data for the Response Variable

Collect the data for the response variables (elasticity) following the order provided in column **RunOrder** in the chosen design. For example, for the CCD design created in step 1, the first condition for the operator corresponding to block 2 to test will be that with component A at level 6.5, component B at level 6.5 and temperature at 60°.

Once all the designed factor level combinations have been tested, enter the recorded response values (elasticity) in the worksheet containing the design. Now you are ready to proceed with the statistical analysis of the collected data.

4.2.2   Plan of the Statistical Analyses

The main focus of a statistical analysis of a *CCD* is to study in depth how several factors contribute to the explanation of a response variable and understand how the factors interact with each other. A second important objective is *optimization*, i.e. the search for those combinations of factor levels that are linked to desirable values of the response.

For these purposes, a second-order model provides a suitable approximation for the unknown functional relationship between factors and response variables. This is useful for identifying regions of factor levels that optimize response variables.

Let's proceed in the following way:

Step 1 – Perform a descriptive analysis (Stat Tool 1.3) of the response.
Step 2 – Fit a second-order model to estimate the effects and determine the significant ones.
Step 3 – If need be, reduce the model to include the significant terms.
Step 4 – Optimize the response.
Step 5 – Examine the shape of the response surface and locate the optimum.

For the Polymer Project, the dataset (File: CCD_Polymer_Project.xlsx) opened in Minitab appears as below.
The variables setting is the following:

- Columns C1 – C3 are related to the previous creation of the CCD design.
- Variables A, B and Temperature are the quantitative factors, each assuming three levels.
- Variable Elasticity is the quantitative response variable.

Column	Variable	Type of data	Label
C4	Blocks	Numeric data	The two operators labeled 1 and 2
C5	A	Numeric data	Additive A with levels: 0.78, 6.5, 12.22
C6	B	Numeric data	Additive B with levels: 0.78, 6.5, 12.22
C7	Temperature	Numeric data	Temperature with levels: 27.3°, 60°, 92.7°
C8	Elasticity	Numeric data	Elasticity

CCD_Polymer_Project.xlsx

4.2.2.1 Step 1 – Perform a Descriptive Analysis of the Response Variables

 As the response is a quantitative variable, let's use the boxplot to describe how the elasticity occurred in our sample, and calculate means and measures of variability to complete the descriptive analysis.

To display the boxplot, go to: **Graph > Boxplot**
Under **One Y** select the graphical option **Simple** and, in the next dialog box, under **Graph variables** select "Elasticity." Then click **OK** in the dialog box.
To change the appearance of a boxplot, use the tips in Chapter 2, section 2.2.2.1, relating to histograms, as well as the following ones:
- To display the boxplot horizontally: double-click on any scale value on the horizontal axis. A dialog box will open with several options to define. Select from these options: Transpose value and category scales.
- To add the mean value to the boxplot: right-click on the area inside the border of the boxplot, then select Add > Data display > Mean symbol.

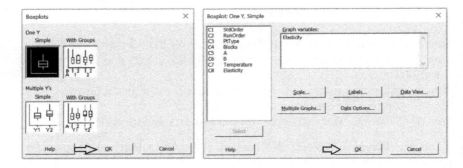

To display the descriptive measures (means, measures of variability, etc.), go to:

Stat > Basic Statistics > Display Descriptive Statistics

In the dialog box, select "Elasticity" under **Variables**, then click **Statistics** to open a dialog box displaying a range of possible statistics to choose from. In addition to the default options, select **Interquartile range** and **Range**.

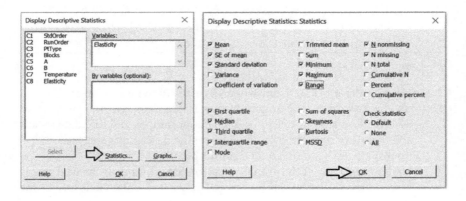

4.2.2.1.1 Interpret the Results of Step 1
Elasticity data is slightly skewed to the left (Stat Tools 1.4–1.5). About 50% of the trials showed an elasticity value greater than 48.9 (the median, Stat Tool 1.6), and around 25% greater than 49.9 (Q_3, Stat Tool 1.7). With respect to *variability*, the average distance of the response values from their mean (standard deviation) is about 1.5 (Stat Tool 1.9).

Boxplot of Elasticity

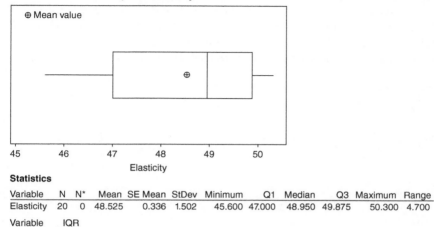

Statistics

Variable	N	N*	Mean	SE Mean	StDev	Minimum	Q1	Median	Q3	Maximum	Range
Elasticity	20	0	48.525	0.336	1.502	45.600	47.000	48.950	49.875	50.300	4.700

Variable	IQR
Elasticity	2.875

4.2.2.2 Step 2 – Fit a Second-Order Model to Estimate the Effects and Determine the Significant Ones

Before analyzing the CCD design, we have to specify the factors within Minitab.

To define the CCD design, go to:

Stat > DOE > Response Surface > Define Custom Response Surface Design

Under **Continuous Factors**, select the factors A, B, and Temperature. If some of your factors are categorical, these should be specified under **Categorical Factors**. Click on **Low/High**. For each factor in the list, verify that the high and low settings are correct, and if these levels are in the original units of the data, select **Uncoded** under **Worksheet data are**. Click **OK**.

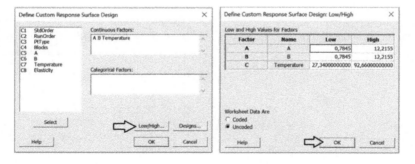

In the main dialog box, click on **Designs**. Specify the columns corresponding to the standard order, run order and point type of the runs. At the bottom of the screen, under Blocks, check the option **Specify by column**, and in the available space select the column representing the blocking variable. Click **OK** in each dialog box. Now you are ready to analyze the response surface design.

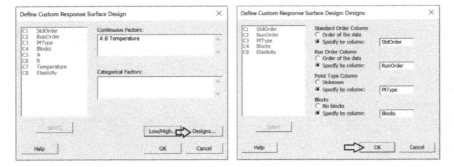

To analyze the response surface design, go to:

Stat > DOE > Response Surface > Analyze Response Surface Design

Select the response variable "Elasticity" in **Responses** and click on **Terms**. At the top of the next dialog box, in **Include terms in the model up through order**, specify **Full quadratic** to study all main effects (both linear and quadratic effects) and the two-way interactions between factors. Select **Include blocks in the model** to consider blocks in the analysis. Click **OK** and in the main dialog box, choose **Graphs**. In **Effects plots** choose **Pareto** and **Normal**. These graphs help you identify the terms (linear and quadratic factor effects and/or interactions) that influence the response and compare the relative magnitude of the effects, along with their statistical significance. Click **OK** in the main dialog box and Minitab shows the results of the analysis, both in the Session window and through the required graphs.

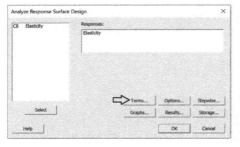

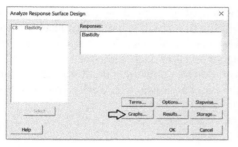

4.2.2.2.1 Interpret the Results of Step 2

First consider the **Pareto chart** that shows which terms contribute the most to explaining the elasticity. Any bar extending beyond the reference line (considering a significance level equal to 5%) represents a significant effect.

In our example, with the exception of the effects involving temperature, the other main effects, the two-factor interactions and the quadratic effects are statistically significant at the 0.05 level.

The Pareto chart shows which effects are statistically significant, but you have no information on how these effects affect the response.

Use the normal plot of the standardized effects to evaluate the direction of the effects.

In the **normal plot of the standardized effects**, the line represents the situation in which all the effects are 0. Effects that are further from 0 and from the line are statistically significant. Minitab shows statistically significant and nonsignificant effects by giving the points different colors and shapes. In addition, the plot indicates the direction of the effect. Positive effects (displayed on the right of the graph) increase the response when the factor moves from its low value to its high value.

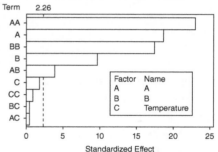

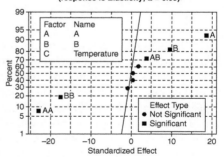

Negative effects (displayed on the left of the graph) decrease the response when moving from the factor's low value to its high value. In our example, the main effects for additives A and B are statistically significant at the 0.05 level. Elasticity seems to increase with higher levels of each additive.

The interaction between the two additives and their quadratic effects are also marked as statistically significant. Later, we will clarify the meaning further.

Response surface regression

ANOVA

Source	DF	Adj SS	Adj MS	F-Value	p-Value
Model	10	42.5324	4.2532	125.46	0.000
Blocks	1	0.3000	0.3000	8.85	0.016
Linear	3	15.0789	5.0263	148.27	0.000
A	1	11.8549	11.8549	349.69	0.000
B	1	3.1197	3.1197	92.02	0.000
Temperature	1	0.1044	0.1044	3.08	0.113
Square	3	26.6435	8.8812	261.98	0.000
A*A	1	17.9803	17.9803	530.38	0.000
B*B	1	10.3571	10.3571	305.51	0.000
Temperature*Temperature	1	0.0231	0.0231	0.68	0.430
Two-way interaction	3	0.5100	0.1700	5.01	0.026
A*B	1	0.5000	0.5000	14.75	0.004
A*Temperature	1	0.0050	0.0050	0.15	0.710
B*Temperature	1	0.0050	0.0050	0.15	0.710
Error	9	0.3051	0.0339		
Lack-of-fit	5	0.2126	0.0425	1.84	0.287
Pure error	4	0.0925	0.0231		
Total	19	42.8375			

In the ANOVA table we examine the p-values to determine whether any factors or interactions, or even blocks, are statistically significant. Remember that the p-value is a probability that measures the evidence against the null hypothesis. Lower probabilities provide stronger evidence against the null hypothesis.

For the main effects, the null hypothesis is that there is no significant difference in mean elasticity across levels of each factor. As we estimate a full quadratic model, we have two tests for each factor that analyze its linear and quadratic effect on response. The statistical significance of the quadratic effects denotes the presence of curvatures in the response surface.

For the two-factor interactions, H_0 states that the relationship between a factor and the response does not depend on the other factor in the term.

We usually consider a significance level alpha equal to 0.05, but in an exploratory phase of the analysis we may also consider a significance level equal to 0.10.

When the p-value is greater than or equal to alpha, we fail to reject the null hypothesis. When it is less than alpha, we reject the null hypothesis and claim statistical significance.

So, if we set the significance level alpha to 0.05, which terms in the model are significant in our example? The answer is: the linear effects of additives A and B; the two-way interactions between these two factors and their quadratic effects. You may now wish to reduce the model to include only significant terms.

When using statistical significance to decide which terms to keep in a model, it is usually advisable not to remove entire groups of nonsignificant terms at the same time. The statistical significance of individual terms can change because of other terms in the model. To reduce your model, you can use an automatic selection procedure, the *stepwise strategy*, to identify a useful subset of terms, choosing one of the three commonly used alternatives (standard stepwise, forward selection, and backward elimination).

4.2.2.3 Step 3 – If Need Be, Reduce the Model to Include the Significant Terms

 To reduce the model, go to:

Stat > DOE > Response Surface > Analyze Response Surface Design

Select the numeric response variable "Elasticity" under **Responses** and click on **Terms**. In **Include the following terms**, specify **Full Quadratic**, to study all main effects (both linear and quadratic effects) and the two-way interactions between factors. Select **Include blocks in the model** to consider blocks in the analysis. Click **OK** and in the main dialog box, choose **Stepwise**. In **Method** select **Backward elimination** and in **Alpha value to remove** specify **0.05**, then click **OK**. In the main dialog box, choose **Graphs**.

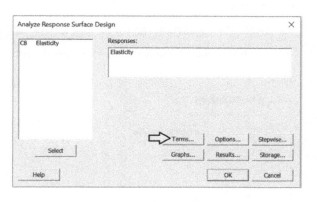

Under **Residual plots** choose **Four in one**. Minitab will display several residual plots to examine whether your model meets the assumptions of the analysis (Stat Tools 2.8, 2.9). Click **OK** in the main dialog box.

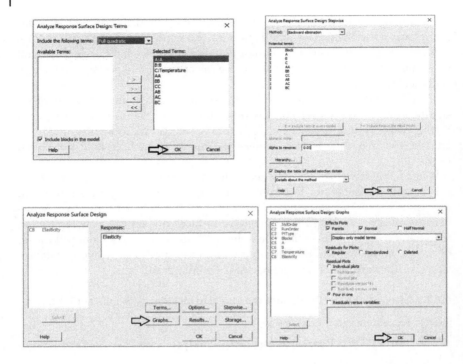

Complete the analysis by adding the factorial plots that show the relationships between the response and the significant terms in the model, displaying how the response mean changes as the factor levels or combinations of factor levels change.

To display factorial plots, go to:

Stat > DOE > Response Surface > Factorial Plots

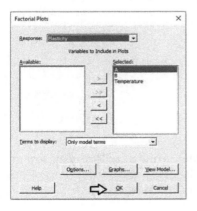

4.2.2.3.1 Interpret the Results of Step 3

By setting the significance level alpha to 0.05, the ANOVA table shows the significant terms in the model. Consider that the stepwise procedure may add nonsignificant terms in order to create a hierarchical model. In a hierarchical model, all lower-order terms that comprise a higher-order term also appear in the model.

Response surface regression

Backward elimination of terms
α to remove = 0.05
ANOVA

Source	DF	Adj SS	Adj MS	F-Value	p-Value
Model	6	42.3949	7.0658	207.53	0.000
Blocks	1	0.3000	0.3000	8.81	0.011
Linear	2	14.9745	7.4873	219.91	0.000
A	1	11.8549	11.8549	348.19	0.000
B	1	3.1197	3.1197	91.63	0.000
Square	2	26.6203	13.3102	390.94	0.000
A*A	1	17.9801	17.9801	528.10	0.000
B*B	1	10.3401	10.3401	303.70	0.000
Two-way interaction	1	0.5000	0.5000	14.69	0.002
A*B	1	0.5000	0.5000	14.69	0.002
Error	13	0.4426	0.0340		
Lack-of-fit	9	0.3501	0.0389	1.68	0.324
Pure error	4	0.0925	0.0231		
Total	19	42.8375			

Model summary

S	R-sq	R-sq(adj)	R-sq(pred)
0.184518	98.97%	98.49%	97.41%

Note that in the ANOVA table, Minitab displays the **lack-of-fit test** when your data contain replicates. This test helps investigators to determine whether the model accurately fits the data. Its null hypothesis states that the model accurately fits the data. As usual, compare the p-value to your significant level α. If the p-value is less than α, reject the null hypothesis and conclude that the lack of fit is statistically significant. If the p-value is larger than α, you fail to reject the null hypothesis and you can conclude that the lack of fit is not statistically significant.

In our example, the p-value (0.324) is greater than 0.05: we conclude that there is no strong evidence of lack of fit. Lack of fit can occur if important terms such as interactions or quadratic terms are not included in the model or if the fitted model shows several, unusually large residuals (Stat Tool 2.9).

In addition to the results of the analysis of variance, Minitab displays some other useful information in the Model Summary table.

The quantity **R-squared** (R-sq, R^2) is interpreted as the percentage of the variability among elasticity data, explained by the terms included in the ANOVA model.

The value of R^2 varies from 0 to 100%, with larger values being more desirable.

The adjusted R^2 (R-sq(adj)) is a variation of the ordinary R^2 that is adjusted for the number of terms in the model. Use adjusted R^2 for more complex experiments with several factors, when you want to compare several models with different numbers of terms.

The value of **S** is a measure of the variability of the errors that we make when we use the ANOVA model to estimate the elasticity data. Generally, the smaller it is, the better the fit of the model to the data.

Before proceeding with the results reported in the Session window, take a look at the residual plots. A **residual** represents an **error**, i.e. the distance between an observed value of the response and its value estimated by the ANOVA model. The graphical analysis of residuals based on **residual plots** helps you to discover possible violations of the underlying ANOVA assumptions (Stat Tools 2.8, 2.9).

A check of the normality assumption could be made looking at the histogram of the residuals (bottom left), but with small samples, often the histogram has an irregular shape.

The normal probability plot of the residuals may be more useful. Here we can see a tendency of the plot to follow the straight line.

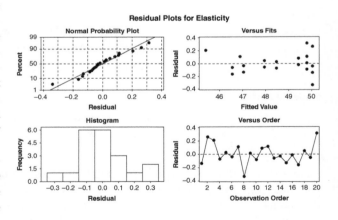

The other two graphs (Residuals versus Fits and Residuals versus Order) seem unstructured, thus supporting the validity of the ANOVA assumptions.

Returning to the Session window, we find a table of coded coefficients followed by a regression equation in uncoded units. In the regression equation, we have a coefficient for each term in the model. They are expressed in

uncoded units, i.e. in the original measurement units. These coefficients are also displayed in the Coded Coefficients table, but in **coded units** (see section 2.2.2.3.1). In the Coded Coefficients table, we find the same results as in the ANOVA table in terms of tests on coefficients and we can evaluate the p-values to determine whether they are statistically significant. The null hypothesis states that there is no effect on the response. Setting the significance level α at 0.05, if a p-value is less than α, reject the null hypothesis and conclude that the effect is statistically significant.

Coded coefficients

Term	Coef	SE Coef	t-Value	p-Value	VIF
Constant blocks	49.9143	0.0648	769.72	0.000	
1	−0.1250	0.0421	−2.97	0.011	1.00
A	1.5398	0.0825	18.66	0.000	1.00
B	0.7899	0.0825	9.57	0.000	1.00
A*A	−3.104	0.135	−22.98	0.000	1.00
B*B	−2.354	0.135	−17.43	0.000	1.00
A*B	0.667	0.174	3.83	0.002	1.00

Regression equation in uncoded units

Elasticity $= 41.069 + 1.3718 \ A + 0.9422 \ B - 0.09501 \ A*A - 0.07205 \ B*B + 0.02041 \ A*B$

Equation averaged over blocks.

Look at the factorial plots to better understand how the mean elasticity varies as factor levels change.

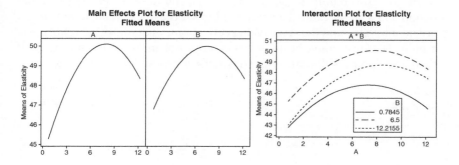

The presence of the quadratic terms A*A and B*B results in a curvature in the factorial plots involving factors A and B. Considering the interaction A*B, the best results in terms of elasticity are obtained when A is around 8 and B is set at its middle value.

4.2.2.4 Step 4 – Optimize the Response

After fitting the response surface model, additional analyses can be carried out to find the optimum set of factor levels that optimizes the response (Stat Tool 3.8).

 To optimize the response variable, go to:

Stat > DOE > Response Surface > Response Optimizer

In the table, under **Goal**, select one of the following options for the response:

- **Do not optimize**: Do not include the response in the optimization process.
- **Minimize**: Lower values of the response are preferable.
- **Target**: The response is optimal when values meet a specific target value.
- **Maximize**: Higher values of the response are preferable.

In our case study the goal is to maximize elasticity: choose **Maximize** from the drop-down list and click **Setup**. In the next window you need to specify a response target and lower or upper bounds depending on your goal. Minitab automatically sets the **Lower value** and the **Upper value** to the smallest and largest values in the data, but you can change these values depending on your response goal (Stat Tool 3.8).

- If your goal is to maximize the response (larger is better), you need to specify the lower bound and a target value. You may want to set the lower value to the smallest acceptable value and the target value at the *point of diminishing returns*, i.e. a point at which going above its value does not make much difference. If there is no point of diminishing returns, use a very high target value.
- If your goal is to target a specific value, you should set the lower value and upper value as points of diminishing returns.
- If the goal is to minimize (smaller is better), you need to specify the upper bound and a target value. You may want to set the upper value to the largest acceptable value and the target value at the point of diminishing returns, i.e. a point at which going below its value does not make much difference. If there is no point of diminishing returns, use a very small target value.

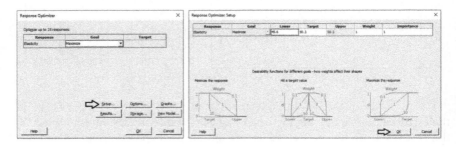

In our example, we leave the default values with the lower bound set to the minimum and the target and upper values set to the maximum.

Note that in the process of response optimization, Minitab calculates and uses desirability indicators from 0 to 1 to represent how well a combination of factor levels satisfies the goals you have for the responses. Specifying a different *weight* varying from 0.1 to 10, you can emphasize or deemphasize the importance of attaining the goals. Furthermore, when you have more than one response, Minitab calculates individual desirabilities for each response and a composite desirability for the entire set of responses. With multiple responses, you may assign a different level of *importance* to each response by specifying how much effect each response has on the composite desirability.

In our example, leave **Weight** and **Importance** equal to 1 and click **OK**. In the main dialog box select **Results** and in **Number of solutions to display**, specify 5. Click **OK** in each dialog box. In the optimization plot, Minitab will display the optimal solution, i.e. the factor setting that maximizes the composite desirability, while in the Session window we will find a list of five different factor combinations close to the optimal one, which is the first on the list.

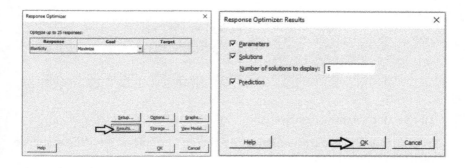

The optimization plot shows how different factor settings affect the response. The vertical lines on the graph represent the current factor settings. The numbers displayed at the top of a column show the current factor level settings (Minitab displays them in red). The horizontal blue lines and numbers represent the responses for the current factor level. The optimal solution is displayed on the graph and serves as the initial point. The settings can then be modified interactively, to determine how different combinations of factor levels affect the response, by moving the red vertical lines corresponding to each factor. To return to the initial settings, right-click and select **Navigation** and then **Reset to Initial Settings**.

4.2.2.4.1 Interpret the Results of Step 4

The optimization plot is a useful tool for exploring the sensitivity of the response variable to changes in the factor settings and in the neighborhood of a local solution – for example, the optimal one.

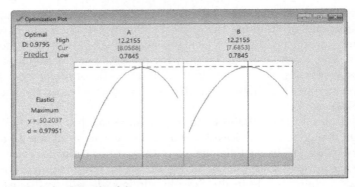

Response Optimization: Elasticity

Parameters

Response	Goal	Lower	Target	Upper	Weight	Importance
Elasticity	Maximum	45.6	50.3		1	1

Solutions

Solution	A	B	Elasticity Fit	Composite Desirability
1	8.0588	7.68525	50.2037	0.979507
2	7.3165	9.23555	49.9554	0.926683
3	11.5136	8.14151	49.0775	0.739901
4	2.5890	6.82780	47.4189	0.386990
5	2.3714	6.71215	47.1908	0.338466

Multiple Response Prediction

Variable	Setting
A	8.05877
B	7.68525

Response	Fit	SE Fit	95% CI	95% PI
Elasticity	50.2037	0.0640	(50.0653; 50.3420)	(49.7817; 50.6256)

The Session window displays:

- Information about the boundaries, weight, and importance for the response variable (Parameters).
- The factor settings, the fitted response, and the composite desirability for the required 5 solutions. The first in the list is the optimal solution (Solutions).
- The factor settings for the optimal solution and the fitted response with their confidence intervals (Stat Tool 1.14) and prediction intervals (Multiple Response Prediction).

For elasticity data the optimal solution has the additive A set to 8.06 and the additive B to 7.68. For this setting, the fitted response (mean elasticity) is equal to 50.20 (95% CI: 50.06-50.34). The composite desirability (0.98) is quite close to 1, denoting a satisfactory overall achievement of the specified goal.

4.2.2.5 Step 5 – Examine the Shape of the Response Surface and Locate the Optimum

Graphic techniques such as *contour, surface,* and *overlaid plots* help us to examine the shape of the response surface in regions of interest. Through these plots, you can explore the potential relationship between *pairs of factors* and the responses. You can display a contour, a surface or an overlaid plot considering two factors at a time, while setting the remaining factors to predefined levels. With contour and surface plots, you consider a single response; in overlaid plots, you can consider more than one response.

To display the contour plot, go to: **Stat > DOE > Response Surface > Contour Plot**
In the dialog box, select the response "Elasticity" in **Response**. Click **Contours**, and under **Data Display**, choose **Contour lines**. Click **OK**.

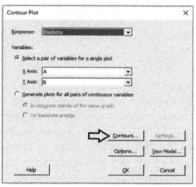

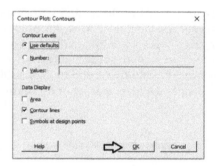

To display the surface plots, go to: **Stat > DOE > Response Surface > Surface Plot**
In the dialog box, select the reponse "Elasticity" in **Response**. Click **OK**.

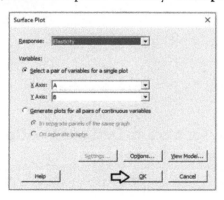

To change the appearance of the surface plot, double-click on any point inside a surface. In **Attributes**, under **Surface Type**, select the option **Wireframe** instead of the default **Surface**.

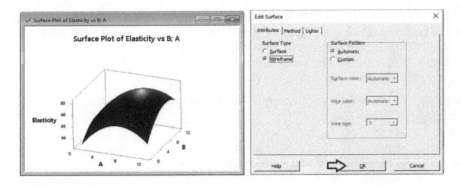

4.2.2.5.1 Interpret the Results of Step 5

In the contour plot, additives A and B are plotted in the x- and y-axes and elasticity is represented by different level curves. Double-click on any point in the frame and select **Crosshairs** to interactively view variations in the response. In the surface plot, elasticity is represented by a smooth surface. It is relatively easy to see that the optimum is very near to 8 for additive A and to 7 for additive B and to explore how the elasticity changes while increasing or decreasing factor levels.

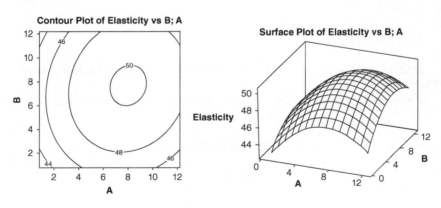

4.3 Case Study for Mixture Designs: Mix-Up Project

In the development of a new detergent, the research team aims to analyze how three different components can affect the product's technical performance.

The three components (A, B, and C) form a mixture of fixed amount. The researchers want to evaluate if changing the proportions of the components inside the mixture will lead to the technical performance (measured as an index assuming values from 0 to 70, with high values representing a better performance) displaying different behavior.

In case study 4.2, we created and analyzed a response surface design in which the levels of each factor can vary independently of the other factors. In this case study, the three components represent ingredients of a mixture and the amount of the mixture is fixed. Consequently, the components are not independent, i.e. if we increase one component, at least one other must decrease. This kind of experiment is called a *mixture experiment* (Stat Tool 4.3).

4.3.1 Plan of the Experimental Design

We will create the experimental design using either a *simplex centroid design* (Stat Tool 4.4) or a *simplex lattice design* (Stat Tool 4.5).

For the present study:

- The three components A, B, and C represent the *key factors*.
- The *technical performance* represents *the response variable*.

To perform the experiment, let's proceed in the following way:

Step 1 – Create a simplex centroid design for a simple mixture experiment.

Step 2 – Alternatively, create a simplex centroid design for a simple mixture experiment with lower limits for components.

Step 3 – Alternatively, create a simplex centroid design for a mixture-process variable experiment.

Step 4 – Alternatively, create a simplex centroid design for a mixture-amount experiment.

Step 5 – Alternatively, create a simplex lattice design for a simple mixture experiment.

Step 6 – Alternatively, create an extreme vertices design with lower and upper limits for components.

Step 7 – Alternatively, create an extreme vertices design with linear constraints for components.

Step 8 – Assign the designed factor level combinations (design points) to the experimental units and collect data for the response variable.

4.3.1.1 Step 1 – Create a Simplex Centroid Design for a Simple Mixture Experiment

Let's start with a simple mixture experiment in which only the proportions of the three components are expected to be related to the technical performance. For this purpose we will create a simplex centroid design, considering the three factors A, B, and C.

To create a simplex centroid design, go to: **Stat > DOE > Mixture > Create Mixture Design**
In **Type of Design** select **Simplex centroid** and in **Number of components** specify 3. Click on **Display Available Designs** to have an indication of the number of runs based on the design type and the number of components in the mixture. Click **OK**.

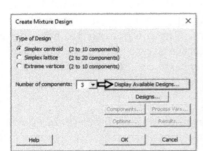
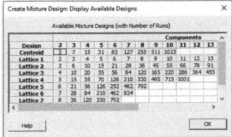

In the main dialog box select **Designs**. Check the option **Augment the design with axial points**. In **Number of replicates for the whole design** specify **1** or, depending on available resources, enter how many times to perform each experimental run. To add replicates for specific design points, for example for a center point or a vertex point, select the option **Number of replicates for the selected types of points**. Remember, you can add replicates to the whole design later with **Stat > DOE > Modify Design**. Click **OK**.

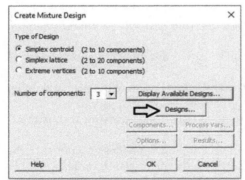
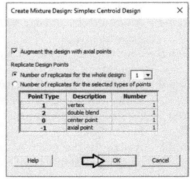

In the main dialog box select **Components**. Check the option **Single total** under **Total Mixture Amount** and leave the default value **1.0** given that the experiment considers a single fixed amount for the mixture and the components are expressed as proportions. Under **Name**, enter a name for each mixture component. For the moment, suppose that each component can vary from

0 to 1. In steps 2 and 6, we will see how to specify lower and/or upper limits for components. Click **OK**.

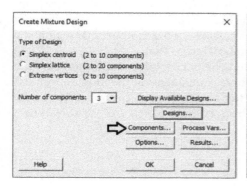
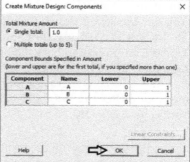

Proceed by clicking **Options** in the main dialog box and check the option **Randomize runs** (Stat Tool 2.3). Click **OK** in each dialog box and Minitab shows the design of the experiment in the worksheet and some information in the Session window.

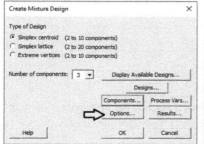

Simplex Centroid Design

Design Summary

Components:	3	Design points:	10
Process variables:	0	Design degree:	3

Mixture total: 1.00000

Number of Boundaries for Each Dimension

Point Type	1	2	0
Dimension	0	1	2
Number	3	3	1

Number of Design Points for Each Type

Point Type	1	2	3	0	-1
Distinct	3	3	0	1	3
Replicates	1	1	0	1	1
Total number	3	3	0	1	3

Bounds of Mixture Components

	Amount		Proportion		Pseudocomponent	
Comp	Lower	Upper	Lower	Upper	Lower	Upper
A	0.0000	1.0000	0.0000	1.0000	0.0000	1.0000
B	0.0000	1.0000	0.0000	1.0000	0.0000	1.0000
C	0.0000	1.0000	0.0000	1.0000	0.0000	1.0000

Worksheet 1 ***

↓	C1	C2	C3	C4	C5	C6	C7	C8
	StdOrder	RunOrder	PtType	Blocks	A	B	C	
1	7	1	0	1	0.33333	0.33333	0.33333	
2	2	2	1	1	0.00000	1.00000	0.00000	
3	4	3	2	1	0.50000	0.50000	0.00000	
4	9	4	-1	1	0.16667	0.66667	0.16667	
5	10	5	-1	1	0.16667	0.16667	0.66667	
6	1	6	1	1	1.00000	0.00000	0.00000	
7	3	7	1	1	0.00000	0.00000	1.00000	
8	8	8	-1	1	0.66667	0.16667	0.16667	
9	6	9	2	1	0.00000	0.50000	0.50000	
10	5	10	2	1	0.50000	0.00000	0.50000	
11								

The design includes 10 design points. In the table, **Number of Design Points for Each Type**, each design point type can be seen. In the design we have three vertices (point type = 1), three double blends (point type = 2), one center point (point type = 0) and three axial points (point type = −1).

You can display the design to graphically check the experimental region.

Go to: **Stat > DOE > Mixture > Simplex Design Plot**

Under **Components**, check the option **Select a triplet of components for a single plot**. If you have more than three components in the mixture, you can select the option **Generate plots for all triplets of components**. Under **Component Unit in Plot(s)**, check **Proportion** and under **Point Labels**, select **Point Type**. Click **OK**.

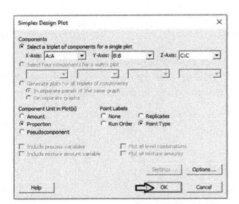

If you hover over the design points, Minitab displays the corresponding mixture. For example, the top vertex corresponds to the pure blend (A = 1, B = 0, C = 0).

Stat Tool 4.3 Mixture Experiments

In a mixture experiment, the factors are ingredients or components of a mixture and these components can't vary independently of each other, i.e. the total amount of the mixture is fixed and if you increase a component, at least one of the other components must decrease. Mixture experiments allow us to study the relationship between mixture components and one or more responses.

Stat Tool 4.3 (Continued)

In mixture experiments, it is important to clarify if you expect the response(s) to be related:

- Only to the mixture of components. In Minitab it is simply a *Mixture experiment*; or
- To the mixture of components and some other factors (process variables) that are not part of the mixture, but might affect its blending properties. For example, the researchers could be interested in evaluating if the temperature is also related to the responses. In Minitab it is a *mixture-process variable experiment* and you can include up to seven process variables. These variables must be binary variables assuming two levels.
- To the mixture of components and the amount of the mixture. For example, you can test a set of mixtures considering two different total amounts: 20 and 30 g. In Minitab, it is a *mixture-amount experiment* and you can specify up to five different total amounts for the mixture.

Regarding the mixture components, they can be considered in the following ways:

- In their original measurement unit, i.e. their individual amount. For example, with two components, say A and B, in the mixtures (15 g, 35 g), (20 g, 30 g), and (0 g, 50 g), the components are expressed in their amounts.
- As proportions of the total amount (generally recommended). For example, we transform the previous mixtures by dividing each component amount by the total amount of the mixture (50 g). In this way the mixtures are shown as (0.3, 0.7), (0.4, 0.6), and (0, 1), with the components now expressed as proportions.

Suppose we have p components expressed as proportions, and each component can vary from 0 to 1. In a mixture experiment we have the following restrictions:

$$0 \le x_i \le 1, i = 1, \ldots, p; x_1 + \ldots + x_p = 1$$

The factor space, i.e. the set of values that components can assume in a mixture experiment, is called the *simplex*. With two components x_1 and x_2, the simplex is a segment line represented in Figure 4.6(a) with two vertices representing mixtures where only one component is present. These vertices represent the so-called *pure blends*. With three components x_1, x_2, and x_3, the simplex is a triangle represented in Figure 4.6(b) with three vertices representing the pure blends. With four components x_1, x_2, x_3, and x_4, the simplex is a tetrahedron represented in Figure 4.6(c) with four vertices representing the pure blends.

Stat Tool 4.3 (Continued)

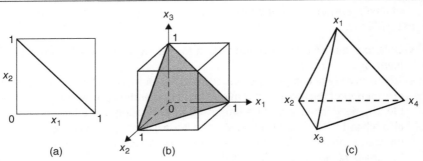

Figure 4.6 The simplex for (a) $p = 2$, (b) $p = 3$, and (c) $p = 4$ mixture components.

In a mixture experiment, the set of mixtures that we can define with the components can be represented as points that lie on the boundaries or inside the simplex.

Creating a *mixture design* means selecting some points on the boundaries or inside the simplex. This is why the so-called *simplex designs* are used to study the effects of mixture components on the response variables.

Statistical software like Minitab can help to make the correct selection. It is therefore important to know how to locate a mixture in the simplex in order to check if the proposed mixture design corresponds adequately to the research needs.

Suppose we have three components in the mixture, expressed as proportions and varying from 0 to 1. A specific mixture will be denoted as (x_1, x_2, x_3). The simplex can be graphically represented by a flat triangle (Figure 4.7), corresponding to the gray oblique triangle shown in Figure 4.6(b).

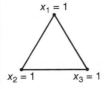

The three vertices correspond to the pure blends or single-component: (1, 0, 0), (0, 1, 0), (0, 0, 1).

Figure 4.7 The simplex for $p = 3$ mixture components.

To locate points inside the simplex, it is useful to draw perpendicular lines from each vertex to the opposite side of the simplex. These lines are called *axial axes*.

Stat Tool 4.3 (Continued)

For example, the axial axis of component x_1 goes from the top to the opposite edge midpoint. The four grid lines mark off 20% increments in the component (Figure 4.8).

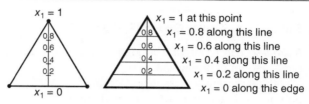

Figure 4.8 Axial axis for component x_1.

By drawing perpendicular lines from a specific point to each axial axis, it is possible to identify the value of the corresponding component in the mixture.

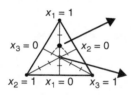

Figure 4.9 Locating design points.

For example, this point inside the simplex corresponds to the mixture (0.6, 0.2, 0.2). Points located inside the simplex correspond to *complete mixtures*, where all the components are present (Figure 4.9).

The intersection point of the three axial axes is called *centroid* and corresponds to the mixture (0.33, 0.33, 0.33), where all the components are present in the same proportion.

Any points along the edges of the simplex represent blends where one of the components is absent (namely the one labeled on the opposite vertex). These are called *two-blend mixtures*.

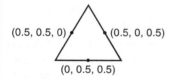

Figure 4.10 Edge midpoints.

The *edge midpoints* are two-blend mixtures in which the two components each make up half of the mixture (Figure 4.10).

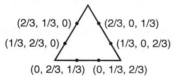

Figure 4.11 Edge trisectors.

The *edge trisectors* are two-blend mixtures in which one component makes up one-third and a second component makes up two-thirds of the mixture (Figure 4.11).

Minitab provides three mixture designs to use according to your needs:

- Simplex centroid designs
- Simplex lattice designs
- Extreme vertices designs

In Stat Tools 4.4–4.7, the main characteristics and purposes of each mixture design are presented.

Stat Tool 4.4 Simplex Centroid Designs

A *simplex centroid design* considers mixtures in which components are of the same amount or proportion.

With three components, expressed as proportions and varying from 0 to 1, the simplex centroid design has the following seven design points (Figure 4.12):

- 3 vertices (pure mixtures)
- 3 two-blend mixtures with two components present in the same proportion
- 1 centroid

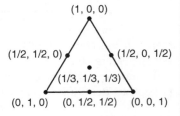

Figure 4.12 Simplex centroid design with $p = 3$ components.

With four components, the simplex centroid design has 15 design points (Figure 4.13):

- 4 vertices (pure mixtures);
- 6 two-blend mixtures with two components present in the same proportion
- 4 three-blend mixtures with three components present in the same proportion
- 1 centroid

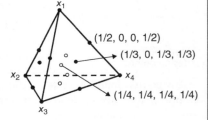

Figure 4.13 Simplex centroid design with $p = 4$ components.

You will notice that in simplex centroid designs, most of the design points are on the boundaries of the simplex (vertices, edges, faces) and there is only one complete mixture represented by the centroid, where all the components are present.

In order to represent more *complete mixtures* inside the simplex, with blends consisting of all mixture components, it is possible to *augment* the basic design with *axial points*.

Axial points are positioned along the axial axes a certain distance from the centroid. Generally, it is recommended that axial points be placed midway between the centroid of the simplex and each vertex.

For example, with three components, expressed as proportions and varying from 0 to 1, the simplex centroid design augmented with axial points has the following 10 design points (Figure 4.14):

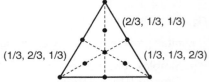

Figure 4.14 Augmented simplex centroid design with $p = 3$ components.

Stat Tool 4.4 (Continued)

- 3 vertices (pure mixtures);
- 3 two-blend mixtures with two components present in the same proportion
- 1 centroid
- 3 three-blend mixtures corresponding to the axial points

Generally, an augmented simplex design is superior when it comes to studying the response of complete mixtures in that it can detect and model curvature within the simplex if there is any in the system. Furthermore, the augmented simplex design has more power to detect lack of fit in the fitted model.

4.3.1.2   Step 2 – Alternatively, Create a Simplex Centroid Design
for a Simple Mixture Experiment with Lower Limits for Components

Suppose you have to specify a *lower limit* for some or all components of the mixture. For example, consider the following constraints on the component proportions: $A \geq 0.05$; $B \geq 0.25$; $C \geq 0.50$. When you specify lower limits to some or all components, the experimental region is still a simplex and you can still create a simplex centroid design.

To create a simplex centroid design with lower limits, go to: **Stat > DOE > Mixture > Create Mixture Design**
Under **Type of Design** select **Simplex centroid** and in **Number of components** specify 3. Select **Designs**. Check the option **Augment the design with axial points**. In **Number of replicates for the whole design** specify **1** or, depending on available resources, enter how many times to perform each experimental run. You can also add replicates for specific design points, for example for a center point or a vertex point, by checking the option **Number of replicates for the selected types of points**. Remember you can add replicates to the whole design later with **Stat > DOE > Modify Design**. Click **OK**.

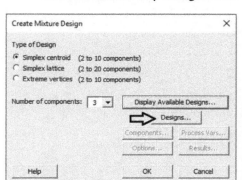

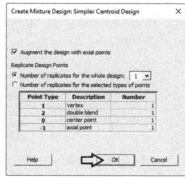

In the main dialog box select **Components**. Check the option **Single total** under **Total Mixture Amount** and leave the default value **1.0** given that the experiment considers a single fixed amount for the mixture and the components are expressed as proportions. Under **Name**, enter a name for each mixture component. Under **Lower**, specify the lower limits for each component. Click **OK**.

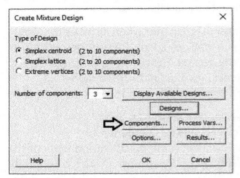
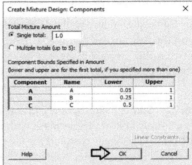

Proceed by clicking **Options** in the main dialog box and check the option **Randomize runs** (Stat Tool 2.3). Click **OK** in each dialog box and Minitab shows the design of the experiment in the worksheet and some information in the Session window.

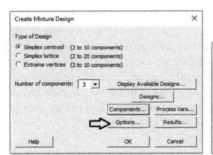
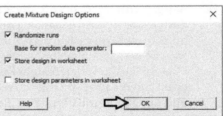

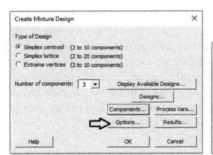

Simplex Centroid Design

Design Summary

Components:	3	Design points:	10
Process variables:	0	Design degree:	3

Mixture total: 1.00000

Number of Boundaries for Each Dimension

Point Type	1	2	0
Dimension	0	1	2
Number	3	3	1

Number of Design Points for Each Type

Point Type	1	2	3	0	-1
Distinct	3	3	0	1	3
Replicates	1	1	0	1	1
Total number	3	3	0	1	3

Bounds of Mixture Components

	Amount		Proportion		Pseudocomponent	
Comp	Lower	Upper	Lower	Upper	Lower	Upper
A	0.050000	0.250000	0.050000	0.250000	0.000000	1.000000
B	0.250000	0.450000	0.250000	0.450000	0.000000	1.000000
C	0.500000	0.700000	0.500000	0.700000	0.000000	1.000000

* NOTE * Bounds were adjusted to accommodate specified constraints.

The design includes 10 design points. The table **Number of Design Points for Each Type** identifies each design point type. Here we have three vertices (point type = 1), three double blends (point type = 2), one center point (point type = 0) and three axial points (point type = −1).

In the table **Bounds of Mixture Components**, you can see that Minitab adjusted the upper limits of each component to accommodate the specified constraints.

You can display the design to graphically check the experimental region.

Go to: **Stat > DOE > Mixture > Simplex Design Plot**

Under **Components**, check the option **Select a triplet of components for a single plot**. If you have more than three components in the mixture, you can select the option **Generate plots for all triplets of components**. Under **Component Unit in Plot(s)**, check **Proportion** and under **Point Labels**, select **Point Type**. Click **OK**.

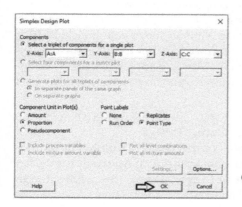
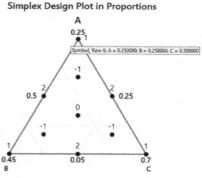

If you hover over the design points, Minitab displays the corresponding mixture. For example, the top vertex corresponds to the complete blend (A = 0.25, B = 0.25, C = 0.50).

You can also display the design in pseudocomponents rather than in proportions.

Go to: **Stat > DOE > Mixture > Simplex Design Plot**

Under **Components**, check the option **Select a triplet of components for a single plot**. If you have more than three components in the mixture, you can select the option **Generate plots for all triplets of components**. Under **Component Unit in Plot(s)**, check **Pseudocomponent** and under **Point Labels**, select **Point Type**. Click **OK**.

If you hover over the design points, Minitab displays the corresponding mixture. Again the top vertex corresponds to the complete blend (A = 0.25, B = 0.25, C = 0.50).

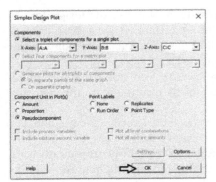

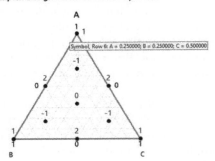

Simplex Design Plot in Pseudocomponents

4.3.1.3 Step 3 – Alternatively, Create a Simplex Centroid Design for a Mixture-Process Variable Experiment

Suppose you now suspect the response is related to the proportions inside the mixture and to other factors which are not part of the mixture. These additional factors are called process variables. Minitab allows you to include up to seven process variables but they must be binary, i.e. factors with only two levels. For example, suppose we need to take into account the possible relationship between temperatures of a washing cycle. For this process variable we set the following levels: 30 and 60 °C.

 To create a simplex centroid design with process variables, go to: **Stat > DOE > Mixture > Create Mixture Design**
In **Type of Design** select **Simplex centroid** and in **Number of components** specify 3. Select **Designs**. Check the option **Augment the design with axial points**. In **Number of replicates for the whole design** specify **1** or, depending on available resources, enter how many times to perform each experimental run. You can also add replicates for specific design points, for example for a center point or a vertex point, by checking the option **Number of replicates for the selected types of points**. Take into account that you can add replicates to the whole design later with **Stat > DOE > Modify Design**. Click **OK**.

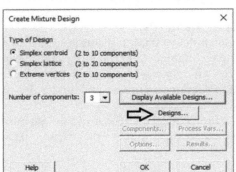

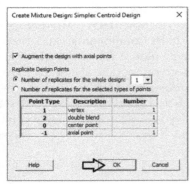

In the main dialog box select **Components**. Check the option **Single total** under **Total Mixture Amount** and leave the default value **1.0** given that the experiment considers a single fixed amount for the mixture and the components are expressed as proportions. Under **Name**, enter a name for each mixture component. Suppose that each component can vary from 0 to 1. You can, however, specify *lower limits* for some or all of the components as in step 2. Click **OK**.

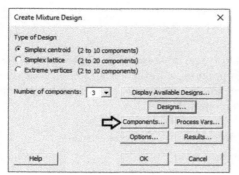
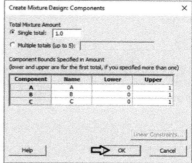

In the main dialog box, select **Process Vars**. Under **Process Variables**, select **Number** and specify **1**. Enter the name of the factor and leave the other default options. Note that Minitab codes the low and high levels of the factor as "–1" and "1."

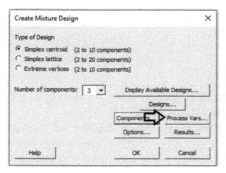
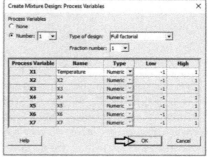

Proceed by clicking **Options** in the main dialog box and check the option **Randomize runs** (Stat Tool 2.3). Click **OK** in each dialog box and Minitab shows the design of the experiment in the worksheet and some information in the Session window.

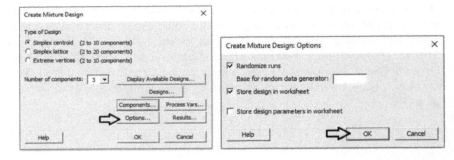

The design includes 20 design points. The table **Number of Design Points for Each Type** identifies each design point type. In the design we have six vertices (point type = 1), six double blends (point type = 2), two center points (point type = 0), and six axial points (point type = −1).

↓	C1 StdOrder	C2 RunOrder	C3 PtType	C4 Blocks	C5 A	C6 B	C7 C	C8 Temperature	C9
1	14	1	2	1	0.50000	0.50000	0.00000	1	
2	2	2	1	1	0.00000	1.00000	0.00000	-1	
3	11	3	1	1	1.00000	0.00000	0.00000	1	
4	16	4	2	1	0.00000	0.50000	0.50000	1	
5	18	5	-1	1	0.66667	0.16667	0.16667	1	
6	1	6	1	1	1.00000	0.00000	0.00000	-1	
7	4	7	2	1	0.50000	0.50000	0.00000	-1	
8	15	8	2	1	0.50000	0.00000	0.50000	1	
9	13	9	1	1	0.00000	0.00000	1.00000	1	
10	12	10	1	1	0.00000	1.00000	0.00000	1	
11	19	11	-1	1	0.16667	0.66667	0.16667	1	
12	5	12	2	1	0.50000	0.00000	0.50000	-1	
13	3	13	1	1	0.00000	0.00000	1.00000	-1	
14	8	14	-1	1	0.66667	0.16667	0.16667	-1	
15	20	15	-1	1	0.16667	0.16667	0.66667	1	
16	9	16	-1	1	0.66667	0.16667	0.16667	-1	
17	6	17	2	1	0.00000	0.50000	0.50000	-1	
18	10	18	-1	1	0.16667	0.16667	0.66667	-1	
19	17	19	0	1	0.33333	0.33333	0.33333	1	
20	7	20	0	1	0.33333	0.33333	0.33333	-1	
21									

Simplex Centroid Design

Design Summary

Components: 3 Design points: 20
Process variables: 1 Design degree: 3

Mixture total: 1.00000

Number of Boundaries for Each Dimension

Point Type	1	2	0
Dimension	0	1	2
Number	3	3	1

Number of Design Points for Each Type

Point Type	1	2	3	0	-1
Distinct	6	6	0	2	6
Replicates	1	1	0	1	1
Total number	6	6	0	2	6

Bounds of Mixture Components

	Amount		Proportion		Pseudocomponent	
Comp	Lower	Upper	Lower	Upper	Lower	Upper
A	0.0000	1.0000	0.0000	1.0000	0.0000	1.0000
B	0.0000	1.0000	0.0000	1.0000	0.0000	1.0000
C	0.0000	1.0000	0.0000	1.0000	0.0000	1.0000

You can display the design to graphically check the experimental region.
Go to: **Stat > DOE > Mixture > Simplex Design Plot**
Under **Components**, check the option **Select a triplet of components for a single plot**. If you have more than three components in the mixture, you can select the option **Generate plots for all triplets of components**. Under **Component Unit in Plot(s)**, check **Proportion**. Check the options **Include process variables** and **Plot all level combinations**. Click **OK**.

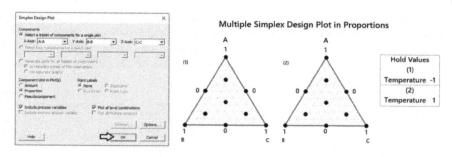

Multiple Simplex Design Plot in Proportions

You can see that the simplex centroid design has been generated at each level of the process variable.

4.3.1.4   Step 4 – Alternatively, Create a Simplex Centroid Design for a Mixture-Amount Experiment

Suppose you now suspect the response is related to both the proportions inside the mixture and the total amount of the mixture. Minitab allows you to specify up to five different amounts. For example, suppose we need to evaluate two different amounts of the mixture equal to 10 and 12 g.

To create a simplex centroid design with multiple mixture amounts, go to: **Stat > DOE > Mixture > Create Mixture Design**
In **Type of Design** select **Simplex centroid** and in **Number of components** specify 3. Select **Designs**. Check the option **Augment the design with axial points**. In **Number of replicates for the whole design** specify **1** or, depending on available resources, enter how many times to perform each experimental run. You can also add replicates for specific design points, for example for a center point or a vertex point, by checking the option **Number of replicates for the selected types of points**. Remember that you can add replicates to the whole design later with **Stat > DOE > Modify Design**. Click **OK**.

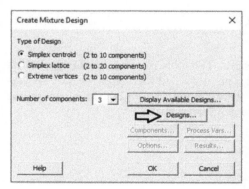

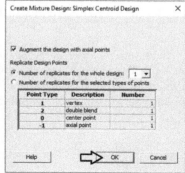

In the main dialog box select **Components**. Check the option **Multiple totals** under **Total Mixture Amount** and specify the two amounts, 10 and 12 g. Under **Name**, enter a name for each mixture component. Note that in the table, the amounts for lower and upper limits must be specified only for the first total (10 g). Click **OK**.

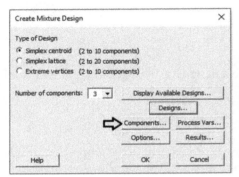
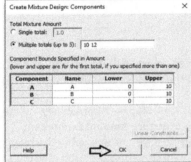

Proceed by clicking **Options** in the main dialog box and check the option **Randomize runs** (Stat Tool 2.3). Click **OK** in each dialog box and Minitab shows the design of the experiment in the worksheet, as well as some information in the Session window.

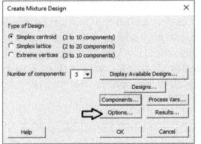
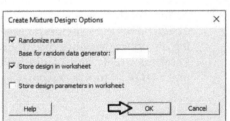

The design includes 20 design points. The table **Number of Design Points for Each Type** identifies each design point type. In the design we have six vertices (point type = 1), six double blends (point type = 2), two center points (point type = 0) and six axial points (point type = −1).

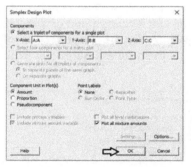

+	C1	C2	C3	C4	C5	C6	C7	C8	C9
	StdOrder	RunOrder	PtType	Blocks	A	B	C	Amount	
1	10	1	-1	1	1.6667	1.6667	6.6667	10	
2	8	2	-1	1	6.6667	1.6667	1.6667	10	
3	12	3	1	1	0.0000	12.0000	0.0000	12	
4	13	4	1	1	0.0000	0.0000	12.0000	12	
5	15	5	2	1	6.0000	0.0000	6.0000	12	
6	9	6	-1	1	1.6667	6.6667	1.6667	10	
7	17	7	0	1	4.0000	4.0000	4.0000	12	
8	19	8	-1	1	2.0000	8.0000	2.0000	12	
9	16	9	2	1	0.0000	6.0000	6.0000	12	
10	11	10	1	1	12.0000	0.0000	0.0000	12	
11	20	11	-1	1	2.0000	2.0000	8.0000	12	
12	14	12	2	1	6.0000	6.0000	0.0000	12	
13	18	13	-1	1	8.0000	2.0000	2.0000	12	
14	3	14	1	1	0.0000	0.0000	10.0000	10	
15	7	15	0	1	3.3333	3.3333	3.3333	10	
16	5	16	2	1	5.0000	0.0000	5.0000	10	
17	4	17	2	1	5.0000	5.0000	0.0000	10	
18	2	18	1	1	0.0000	10.0000	0.0000	10	
19	1	19	1	1	10.0000	0.0000	0.0000	10	
20	6	20	2	1	0.0000	5.0000	5.0000	10	
21									

Simplex Centroid Design

Design Summary

Components:	3	Design points:	20
Process variables:	0	Design degree:	3

Mixture totals: 10.00000; 12.00000

Number of Boundaries for Each Dimension

Point Type	1	2	0
Dimension	0	1	2
Number	3	3	1

Number of Design Points for Each Type

Point Type	1	2	3	0	-1
Distinct	6	6	0	2	6
Replicates	1	1	0	1	1
Total number	6	6	0	2	6

Bounds of Mixture Components (first amount)

Comp	Amount		Proportion		Pseudocomponent	
	Lower	Upper	Lower	Upper	Lower	Upper
A	0.000	10.000	0.0000	1.0000	0.0000	1.0000
B	0.000	10.000	0.0000	1.0000	0.0000	1.0000
C	0.000	10.000	0.0000	1.0000	0.0000	1.0000

You can display the design to graphically check the experimental region.

Go to: **Stat > DOE > Mixture > Simplex Design Plot**

Under **Components**, check the option **Select a triplet of components for a single plot**. If you have more than three components in the mixture, you can select the option **Generate plots for all triplets of components**. Under **Component Unit in Plot(s)**, check **Amount** (or **Proportion** if you want to display the plot in proportions). Check the options **Plot all mixture amounts**. Click **OK**.

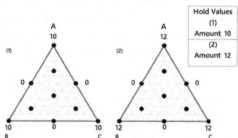

You can see that the simplex centroid design has been generated at each level of the total amount of the mixture.

4.3.1.5 Step 5 – Alternatively, Create a Simplex Lattice Design for a Simple Mixture Experiment

Let us again consider a simple mixture experiment in which only the proportions of the three components are expected to be related to technical performance. For this purpose we will now create a simplex lattice design (Stat Tool 4.5), considering the three factors A, B, and C.

To create a simplex lattice design, go to: **Stat > DOE > Mixture > Create Mixture Design**

In **Type of Design** select **Simplex lattice** and in **Number of components** specify 3. Click on **Display Available Designs** to get an idea of the number of runs based on the design type, the number of components in the mixture and the degree of the lattice. Click **OK**.

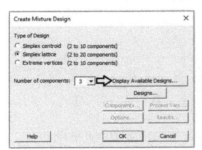
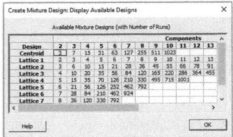

In the main dialog box select **Designs**. In **Degree of lattice** specify for example **3** in order to include the *edge trisectors* that are two-blend mixtures in which one component makes up one-third and a different component makes up two-thirds of the mixture. Check the options **Augment the design with center point** and **Augment the design with axial points**. In **Number of replicates for the whole design** specify **1** or, depending on available resources, enter how many times to perform each experimental run. You can also add replicates for specific design points, for example for a center point or a vertex point, by checking the option **Number of replicates for the selected types of points**. Remember you can add replicates to the whole design later with **Stat > DOE > Modify Design**. Click **OK**.

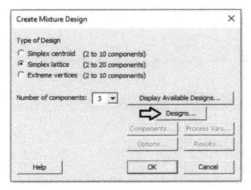
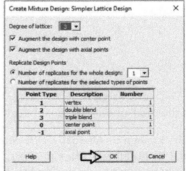

In the main dialog box select **Components**. Check the option **Single total** under **Total Mixture Amount** and leave the default value **1.0**, given that the experiment considers a single fixed amount for the mixture and the components are expressed as proportions. Under **Name**, enter a name for each mixture component. Suppose that each component can vary from 0 to 1. You can, however, specify lower limits for some or all of the components. Click **OK**.

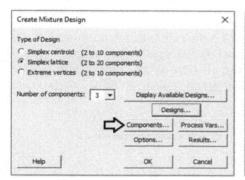
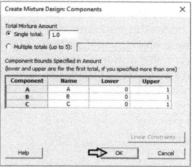

Proceed by clicking **Options** in the main dialog box and check the option **Randomize runs** (Stat Tool 2.3). Click **OK** in each dialog box and Minitab shows the design of the experiment in the worksheet, as well as some information in the Session window.

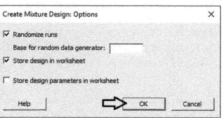

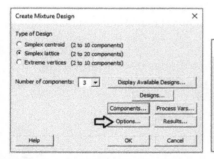

Simplex Lattice Design
Design Summary

Components:	3	Design points:	13
Process variables:	0	Lattice degree:	3

Mixture total: 1.00000

Number of Boundaries for Each Dimension

Point Type	1	2	0
Dimension	0	1	2
Number	3	3	1

Number of Design Points for Each Type

Point Type	1	2	3	0	-1
Distinct	3	6	0	1	3
Replicates	1	1	0	1	1
Total number	3	6	0	1	3

Bounds of Mixture Components

Comp	Amount Lower	Amount Upper	Proportion Lower	Proportion Upper	Pseudocomponent Lower	Pseudocomponent Upper
A	0.0000	1.0000	0.0000	1.0000	0.0000	1.0000
B	0.0000	1.0000	0.0000	1.0000	0.0000	1.0000
C	0.0000	1.0000	0.0000	1.0000	0.0000	1.0000

The design includes 13 design points. The table **Number of Design Points for Each Type** identifies each design point type. In the design we have three vertices (point type = 1), six double blends (point type = 2), one center point (point type = 0) and three axial points (point type = −1).

You can display the design to graphically check the experimental region.

Go to: **Stat > DOE > Mixture > Simplex Design Plot**

Under **Components**, check the option **Select a triplet of components for a single plot**. If you have more than three components in the mixture, you can select the option **Generate plots for all triplets of components**. Under **Component Unit in Plot(s)**, check **Proportion** and under **Point Labels**, select **Point Type**. Click **OK**.

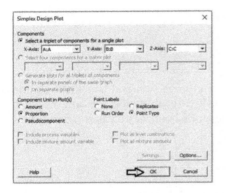

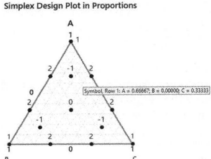

If you hover over the design points, Minitab displays the corresponding mixture. For example, the edge trisector indicated on the right corresponds to the two-blend mixture (A = 0.67, B = 0, C = 0.33).

Stat Tool 4.5 Simplex Lattice Designs

A *simplex lattice design* considers design points arranged in a uniform way (lattice) in the simplex. For this kind of mixture designs we have to specify a parameter called the *degree* of the lattice. In a simplex lattice design for p components with a degree equal to m, the proportions assumed by each component take the $m + 1$ equally spaced values from 0 to 1:

$$x_i = 0, 1/m, 2/m, \ldots, 1 \quad i = 1, 2, \ldots, p.$$

For example, with three components in the simplex lattice design of degree 3, we have:

$$x_i = 0, 1/3, 2/3, 1 \quad i = 1, 2, 3$$

Stat Tool 4.5 (Continued)

The design will consist of the following 10 design points (Figure 4.15):

- 3 vertices (pure mixtures);
- 6 two-blend mixtures corresponding to the edge trisectors;
- 1 centroid

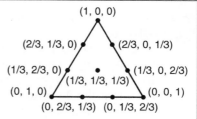

Figure 4.15 Simplex lattice design of degree 3 with $p = 3$ components.

Note that for some degrees the simplex lattice design does not include the centroid. In these cases it is possible to augment the basic design with the *centroid*.

Additionally, in order to have more complete mixtures inside the simplex, where the blends will consist of all mixture components, it is possible to *augment* the basic design with *axial points*.

The axial points are positioned along the axial axes a certain distance from the centroid. Generally, it is recommended that axial runs be placed midway between the centroid of the simplex and each vertex.

For example (Figure 4.16), with three components, expressed as proportions and varying from 0 to 1, the simplex lattice design of degree 3 augmented with axial points has the following 13 design points:

- 3 vertices (pure mixtures);
- 6 two-blend mixtures corresponding to the edge trisectors;
- 1 centroid;
- 3 three-blend mixtures corresponding to the axial points

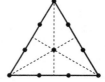

Figure 4.16 Augmented simplex lattice design of degree 3 with $p = 3$ components.

Generally, an augmented simplex design is superior when it comes to studying the response of complete mixtures in that it can detect and model curvature within the simplex if there is any in the system. Furthermore, the augmented simplex design has more power to detect lack of fit in the fitted model.

4.3.1.6 Step 6 – Alternatively, Create an Extreme Vertices Design with Lower and Upper Limits for Components

Suppose that you have to specify *lower and upper limits* or only *upper limits* for some or all components of the mixture. For example, consider the following constraints on the component proportions: $0.05 \leq A \leq 0.25$; $0.25 \leq B \leq 0.40$; $0.50 \leq C \leq 0.70$. In this case the experimental region is no longer a simplex and we can use an extreme vertices design (Stat Tool 4.7). Furthermore, suppose that we want to perform a maximum of 14 runs including some replicates.

 To create an extreme vertices design, go to: **Stat** > **DOE** > **Mixture** > **Create Mixture Design**
In **Type of Design** select **Extreme vertices** and in **Number of components** specify 3. Select **Designs**. In **Degree of design** specify for example 2. Check the options **Augment the design with center point** and **Augment the design with axial points**. In **Number of replicates for the whole design** specify 1. Click **OK**.

Stat Tool 4.6 Constrained Simplex Designs

Until now we have considered mixture components expressed in proportions and varying from 0 to 1. Sometimes the researcher needs to specify *lower bounds* when any of the components must be in the mixture or *upper bounds* when the mixture cannot contain more than a specified proportion of an ingredient. In these cases the mixture design is said to be *constrained*.

If the researcher specifies *only* lower limits for some or all mixture components, the factor space is still a *simplex*, but it is inscribed inside the original simplex region.

For example, Figure 4.17 shows the simplex (the gray triangle) defined by specifying the following limits for three mixture components: $x_1 \geq 0.1$, $x_2 \geq 0.2$, $x_3 \geq 0.3$. Note that with lower limits, the upper limits are rearranged and it is also possible to express the mixture components as *pseudocomponents* denoted by $x'_1, \ldots, x'_p$.

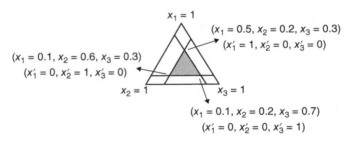

Figure 4.17 Constrained mixture design.

With the specification of lower limits, it is still possible to create a simplex design (centroid or lattice) because the constrained region is still a simplex.

With upper limits or *both* lower and upper limits for at least one of the components, the factor space is no longer a simplex. In these cases, *extreme vertices designs* (Stat Tool 4.7) can be used to select design points in the irregular factor space.

Stat Tool 4.7 Extreme Vertices Designs

In constrained designs (Stat Tool 4.6), the presence of *upper bound constraints* or both *lower and upper bound constraints* on the mixture components and/or the specification of *linear constraints* create a factor space that is no longer a simplex.

With this kind of constraints, you can't create simplex centroid or lattice designs. Instead, extreme vertices designs can be used for this type of mixture problem. Extreme vertices designs cover only a subportion or smaller space within the simplex to take into account component bounds and /or linear constraints.

Notice that when you specify lower and upper limits on:

- An *individual component*, you are setting a *component bound*, for example: $0.2 \leq x_1 \leq 0.6$.
- A *set of components*, you are setting a *linear constraint*, for example: $0.3 \leq x_1 + x_2 \leq 0.6$. Linear constraints can also be of the type: $x_1 \leq x_2$.

For example, Figure 4.18 shows the factor space (the gray region) obtained by specifying the following component bounds: $0.1 \leq x_1 \leq 0.6$.

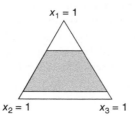

$$x_1 = 1$$

$$x_2 = 1 \qquad x_3 = 1$$

Figure 4.18 Constrained mixture design.

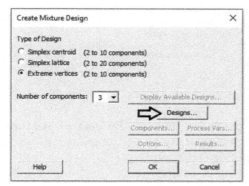

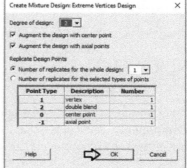

In the main dialog box select **Components**. Check the option **Single total** under **Total Mixture Amount** and leave the default value **1.0**, given that the experiment considers a single fixed amount for the mixture and the components are expressed as proportions. Under **Name**, enter a name for each mixture component. Under **Lower** and **Upper**, specify the lower and upper limits for each component. Click **OK**.

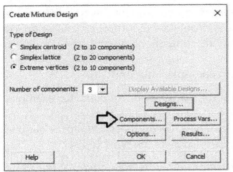
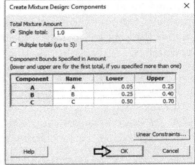

Proceed by clicking **Options** in the main dialog box and uncheck the option **Randomize runs** since for the moment, we just want to check the design. Then, when we have decided that the design is suitable for our needs, we will randomize the runs. Click **OK** in each dialog box and Minitab shows the design of the experiment in the worksheet, as well as some information in the Session window.

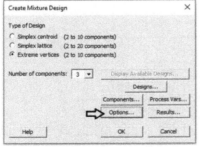
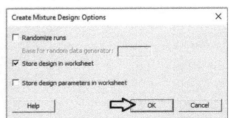

The design includes 13 design points. In the table **Bounds of Mixture Components**, you can see that Minitab created a pseudocomponent for each component to accommodate the specified constraints.

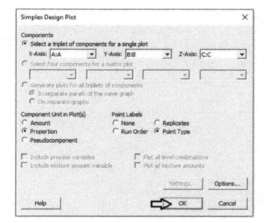

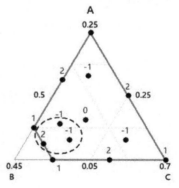

Extreme Vertices Design

Design Summary

Components:	3	Design points:	13
Process variables:	0	Design degree:	2

Mixture total: 1.00000

Number of Boundaries for Each Dimension

Point Type	1	2	0
Dimension	0	1	2
Number	4	4	1

Number of Design Points for Each Type

Point Type	1	2	3	0	-1
Distinct	4	4	0	1	4
Replicates	1	1	0	1	1
Total number	4	4	0	1	4

Bounds of Mixture Components

	Amount		Proportion		Pseudocomponent	
Comp	Lower	Upper	Lower	Upper	Lower	Upper
A	0.050000	0.250000	0.050000	0.250000	0.000000	1.000000
B	0.250000	0.400000	0.250000	0.400000	0.000000	0.750000
C	0.500000	0.700000	0.500000	0.700000	0.000000	1.000000

You can display the design to graphically check the experimental region.

Go to: **Stat** > **DOE** > **Mixture** > **Simplex Design Plot**

Under **Components**, check the option **Select a triplet of components for a single plot**. If you have more than three components in the mixture, you can select the option **Generate plots for all triplets of components**. Under **Component Unit in Plot(s)**, check **Proportion** (or **Pseudocomponent**) and under **Point Labels**, select **Point Type**. Click **OK**.

You can see that the region is not a simplex. Furthermore, you see that three design points inside the marked dotted oval are very close to each other and to other design points nearby. Given that we have to create a design with 14 design points that includes some replicates, we can try to select an *optimal design* from the basic extreme vertices design. An *optimal* design represents a design that is *best* according to some statistical criterion. It includes only a *suitable subset* of all combinations of factors (mixtures of components) in order to reduce the number of runs. For this reason, they are often used when there are some constraints in the number of runs to perform.

In our example, it would be desirable to keep all the design points of the basic design except the three points inside the dotted oval and then to augment the design with some replicates to obtain an optimal design with 14 runs. With Minitab, you can create an indicator column that identifies any experimental run that must be in the optimal design. This indicator column specifies a number for each design point. The sign of the number identifies whether an experimental run must be in the optimal design: a positive sign identifies a point that can be left out of the design, while a negative sign denotes a point that must remain in the optimal design.

To better identify the three design points that can be left out of the design, let's again create the simplex plot by displaying the run order of each design point.

Go to: **Stat > DOE > Mixture > Simplex Design Plot**

Under **Components**, check the option **Select a triplet of components for a single plot**. Under **Component Unit in Plot(s)**, check **Proportion** (or **Pseudocomponent**) and under **Point Labels**, select **Run Order**. Click OK.

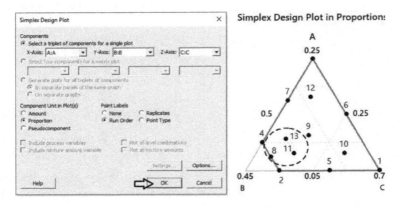

You can see that the three design points that can be left out of the design, have Run Order 8, 11, and 13 in the worksheet of the basic design.

Let's create the indicator column specifying "1" for the three points with Run Order 8, 11, and 13, and "−1" for all the remaining points. The worksheet will appear in the following way:

	C1	C2	C3	C4	C5	C6	C7	C8	C9
	StdOrder	RunOrder	PtType	Blocks	A	B	C	Indicator_column	
2	2	2	1	1	0.05000	0.4000	0.55000	-1	
3	3	3	1	1	0.25000	0.2500	0.50000	-1	
4	4	4	1	1	0.10000	0.4000	0.50000	-1	
5	5	5	2	1	0.05000	0.3250	0.62500	-1	
6	6	6	2	1	0.15000	0.2500	0.60000	-1	
7	7	7	2	1	0.17500	0.3250	0.50000	-1	
8	8	8	2	1	0.07500	0.4000	0.52500	1	
9	9	9	0	1	0.11250	0.3250	0.56250	-1	
10	10	10	-1	1	0.08125	0.2875	0.63125	-1	
11	11	11	-1	1	0.08125	0.3625	0.55625	1	
12	12	12	-1	1	0.18125	0.2875	0.53125	-1	
13	13	13	-1	1	0.10625	0.3625	0.53125	1	
14									

To select the optimal design, go to: **Stat > DOE > Mixture > Select Optimal Design**

Under **Criterion**, check the option **D-Optimality**. Under **Task**, choose **Augment/improve design** and in the blank field select the indicator column. In **Number of points in optimal design**, specify 14. Click on **Terms**. In the next dialog box, beside **Include component terms for model**, specify **Quadratic** to detect curvatures if present. Click **OK** in each dialog box and Minitab shows the optimal design in a new worksheet, as well as some information in the Session window.

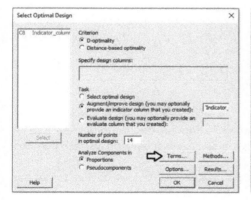
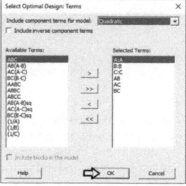

↕	C1	C2	C3	C4	C5	C6	C7	C8
	StdOrder	RunOrder	PtType	Blocks	A	B	C	
1	1	1	1	1	0.05000	0.2500	0.70000	
2	2	2	1	1	0.05000	0.4000	0.55000	
3	3	3	1	1	0.25000	0.2500	0.50000	
4	4	4	1	1	0.10000	0.4000	0.50000	
5	5	5	2	1	0.05000	0.3250	0.62500	
6	6	6	2	1	0.15000	0.2500	0.60000	
7	7	7	2	1	0.17500	0.3250	0.50000	
8	9	9	0	1	0.11250	0.3250	0.56250	
9	10	10	-1	1	0.08125	0.2875	0.63125	
10	12	12	-1	1	0.18125	0.2875	0.53125	
11	1	1	1	1	0.05000	0.2500	0.70000	
12	3	3	1	1	0.25000	0.2500	0.50000	
13	6	6	2	1	0.15000	0.2500	0.60000	
14	5	5	2	1	0.05000	0.3250	0.62500	
15								

You can display the design to graphically check the experimental region.

Go to: **Stat > DOE > Mixture > Simplex Design Plot**

Under **Components,** check the option **Select a triplet of components for a single plot.** Under **Component Unit in Plot(s),** check **Proportion (or Pseudocomponent)** and under **Point Labels,** select **Replicates.** Click **OK.**

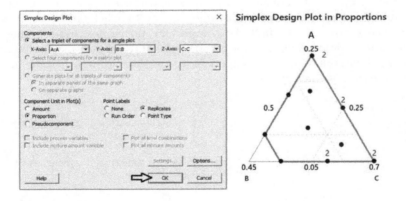

You can see that the design includes 10 different design points and four of these runs (two vertices and two edge centers) are replicated and are denoted in the simplex plot by the number "2."

The last step is to randomize the runs. Before this step, we have to renumber the runs.

Go to: **Stat > DOE > Modify Design**

Under **Modification,** check the option **Renumber design.** Click **OK.**

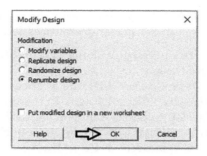

Finally to randomize the runs, go to: **Stat > DOE > Modify Design**
Under **Modification**, check the option **Randomize design**. Select **Specify**
and then click OK in each dialog box.

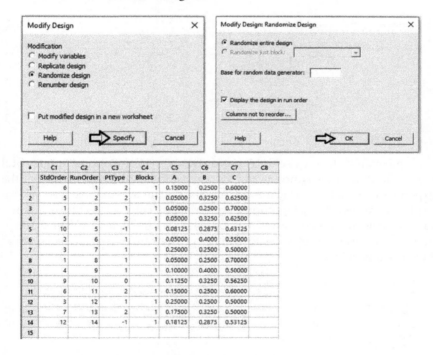

	C1	C2	C3	C4	C5	C6	C7	C8
	StdOrder	RunOrder	PtType	Blocks	A	B	C	
1	6	1	2	1	0.15000	0.2500	0.60000	
2	5	2	2	1	0.05000	0.3250	0.62500	
3	1	3	1	1	0.05000	0.2500	0.70000	
4	5	4	2	1	0.05000	0.3250	0.62500	
5	10	5	-1	1	0.08125	0.2875	0.63125	
6	2	6	1	1	0.05000	0.4000	0.55000	
7	3	7	1	1	0.25000	0.2500	0.50000	
8	1	8	1	1	0.05000	0.2500	0.70000	
9	4	9	1	1	0.10000	0.4000	0.50000	
10	9	10	0	1	0.11250	0.3250	0.56250	
11	6	11	2	1	0.15000	0.2500	0.60000	
12	3	12	1	1	0.25000	0.2500	0.50000	
13	7	13	2	1	0.17500	0.3250	0.50000	
14	12	14	-1	1	0.18125	0.2875	0.53125	
15								

4.3.1.7 Step 7 – Alternatively, Create an Extreme Vertices Design
with Linear Constraints for Components

Suppose that you have to specify one or more *linear constraints* for some or all
components of the mixture. For example, suppose that the sum of the proportions of components A and B in the mixture must be between 0.3 and 0.6
$(0.3 \leq A + B \leq 0.6)$ and that component C must always be no less than the

proportion of component A ($C \geq B$). In this case, the experimental region is no longer a simplex and we can use an extreme vertices design (Stat Tool 4.7).

 To create an extreme vertices design, go to: **Stat > DOE > Mixture > Create Mixture Design**
In **Type of Design** select **Extreme vertices** and in **Number of components** specify 3. Select **Designs**. In **Degree of design** specify for example **2**. Check the options **Augment the design with center point** and **Augment the design with axial points**. In **Number of replicates for the whole design** specify 1. Click **OK**.

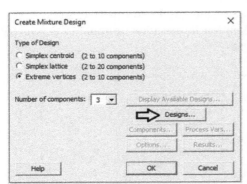
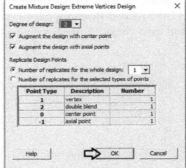

In the main dialog box select **Components**. Check the option **Single total** under **Total Mixture Amount** and leave the default value **1.0**, given that the experiment considers a single fixed amount for the mixture and the components are expressed as proportions. Under **Name**, enter a name for each mixture component. Under **Lower** and **Upper**, leave the default values 0 and 1. You can, however, specify lower and/or upper limits for the components. Select **Linear Constraints**.

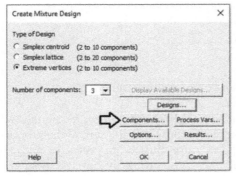
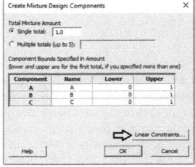

Enter the linear constraints by specifying a coefficient for each component of the mixture and a value for Lower and/or Upper, then click **OK**.

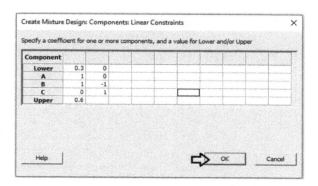

Proceed by clicking **Options** in the main dialog box and check the option **Randomize runs**. Click **OK** in each dialog box and Minitab shows the design of the experiment in the worksheet, as well as some information in the Session window.

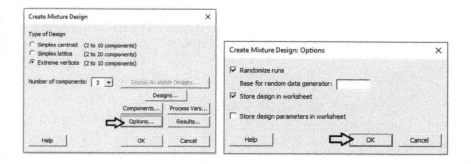

The design includes 16 design points. In the table **Bounds of Mixture Components**, you can see that Minitab created a pseudocomponent for each component to accommodate the specified constraints.

Worksheet 1 ***

+	C1 StdOrder	C2 RunOrder	C3 PtType	C4 Blocks	C5 A	C6 B	C7 C	C8
1	15	1	-1	1	0.11	0.37	0.52	
2	16	2	-1	1	0.21	0.32	0.47	
3	13	3	-1	1	0.11	0.27	0.62	
4	1	4	1	1	0.30	0.00	0.70	
5	8	5	2	1	0.15	0.15	0.70	
6	12	6	-1	1	0.26	0.12	0.62	
7	2	7	1	1	0.00	0.30	0.70	
8	7	8	2	1	0.45	0.00	0.55	
9	3	9	1	1	0.60	0.00	0.40	
10	10	10	2	1	0.10	0.45	0.45	
11	6	11	2	1	0.00	0.40	0.60	
12	14	12	-1	1	0.41	0.12	0.47	
13	9	13	2	1	0.40	0.20	0.40	
14	4	14	1	1	0.00	0.50	0.50	
15	5	15	1	1	0.20	0.40	0.40	
16	11	16	0	1	0.22	0.24	0.54	
17								

Extreme Vertices Design

Design Summary

Components: 3   Design points: 16
Process variables: 0   Design degree: 2

Mixture total: 1.00000

Number of Boundaries for Each Dimension

Point Type 1 2 0
Dimension 0 1 2
Number 5 5 1

Number of Design Points for Each Type

Point Type 1 2 3 0 -1
Distinct 5 5 0 1 5
Replicates 1 1 0 1 1
Total number 5 5 0 1 5

Bounds of Mixture Components

	Amount		Proportion		Pseudocomponent	
Comp	Lower	Upper	Lower	Upper	Lower	Upper
A	0.00000	0.60000	0.00000	0.60000	0.00000	1.00000
B	0.00000	0.50000	0.00000	0.50000	0.00000	0.83333
C	0.40000	0.70000	0.40000	0.70000	0.00000	0.50000

* NOTE * Bounds were adjusted to accommodate specified constraints.

Linear Constraints of Mixture Components

Constraint	Lower	A	B	C	Upper
1	0.30000	1.00000	1.00000	0.00000	0.60000
2	0.00000	0.00000	-1.00000	1.00000	

You can display the design to graphically check the experimental region.
Go to: **Stat > DOE > Mixture > Simplex Design Plot**
Under **Components**, check the option **Select a triplet of components for a single plot**. If you have more than three components in the mixture, you can select the option **Generate plots for all triplets of components**. Under **Component Unit in Plot(s)**, check **Proportion** (or **Pseudocomponent**) and under **Point Labels**, select **Point Type**. Click **OK**.

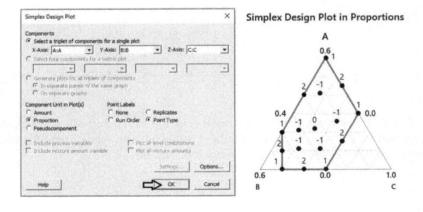

You can see that the region is not a simplex. If required you can decide to select an optimal design from the basic one as done previously in step 6.

4.3.1.8 Step 8 – Assign the Designed Factor Level Combinations (Design Points) to the Experimental Units and Collect Data for the Response Variable

Collect the data for the response variable, i.e. the technical performance measured on a scale of 0 to 70, following the order given in the column **RunOrder** in the chosen design. For example, for the extreme vertices design created in step 6, the first mixture to test will be the mixture with component A at level 0.15, component B at 0.25 and C at 0.60, where components are expressed as proportions.

Once all the designed mixtures have been tested, enter the recorded response values (technical performance index) in the worksheet containing the design. Now you are ready to proceed with the statistical analysis of the collected data.

Stat Tool 4.8 Mixture Models

In Mixture experiments, you can model linear or curvilinear aspects of the response surface by fitting models with different orders. Figure 4.19 shows an example where the response surface is fitted by a first-order (linear) model (a) and a second-order (quadratic) model (b) in a mixture experiment with three components. You can see that the quadratic model allows the detection of curvatures in the response surface if present in the system.

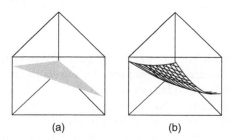

(a)   (b)

Figure 4.19 Response surface plots for a mixture experiment with three components.

4.3.2   Plan of the Statistical Analyses

Second-order models can represent a suitable approximation for the unknown functional relationship to study the effects of mixture components on the response variable.

Let's proceed in the following way:

Step 1 – Perform a descriptive analysis (Stat Tool 1.3) of the response variables.

Step 2 – Fit a second-order model to estimate the effects and determine the significant ones.

Step 3 – Optimize the response.

Step 4 – Examine the shape of the response surface and locate the optimum.

For the Mix-Up Project, the dataset (File: Mix_Up_Project.xlsx) opened in Minitab appears as below.

The variables setting is the following:

- Columns C1–C4 are related to the previous creation of the extreme vertices design.
- Variables A, B, and C are the quantitative mixture components.
- Variable Tech_Index is the quantitative response variable.

Column	Variable	Type of data	Label
C5	A	Numeric data	Proportion of component A of the mixture
C6	B	Numeric data	Proportion of component B of the mixture
C7	C	Numeric data	Proportion of component C of the mixture
C8	Tech_Index	Numeric data	Technical performance index

Mix_Up_Project.xlsx

4.3.2.1   Step 1 – Perform a Descriptive Analysis of the Response Variables

As the response is a quantitative variable, let's use a boxplot to describe how the technical performance index occurred in our sample, and calculate means and measures of variability to complete the descriptive analysis.

 To display the boxplot, go to: **Graph** > **Boxplot**

Under **One Y**, check the graphical option **Simple** and in the next dialog box, select "Tech_Index" under **Graph variables**. Then click **OK** in the dialog box.

To change the appearance of a boxplot, use the tips in Chapter 2, Section 2.2.2.1 to histograms, as well as the following ones:

- To display the boxplot horizontally: double-click on any scale value on the horizontal axis. A dialog box will open with several options to define. Select from these options: Transpose value and category scales.

- To add the mean value to the boxplot: right-click in the area inside the border of the boxplot and select Add > Data display > Mean symbol.

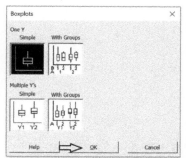
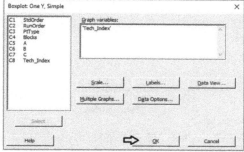

To display the descriptive measures (means, measures of variability, etc.), go to:

Stat > Basic Statistics > Display Descriptive Statistics

In the dialog box, select "Tech_Index" in **Variables**, then click on **Statistics** to open a dialog box displaying a range of possible statistics to choose from. In addition to the default options, select **Interquartile range** and **Range**.

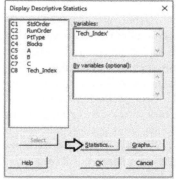
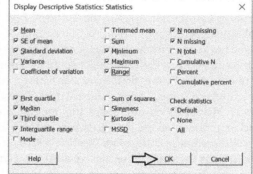

4.3.2.1.1 Interpret the Results of Step 1

Technical performance data are slightly skewed with two outliers showing high values of performance. About 50% of the trials showed a performance score greater than 29.9 (the median, Stat Tool 1.6), and around 25%, greater than 34.8 (Q₃, Stat Tool 1.7). With respect to *variability*, the average distance of the response values from their mean (standard deviation) is about 15.9 (Stat Tool 1.9).

Boxlot of Tech_Index

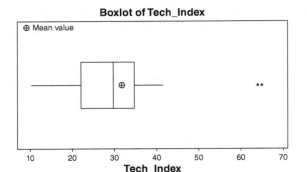

Statistics

Variable	N	N*	Mean	SE Mean	StDev	Minimum	Q1	Median	Q3	Maximum	Range
Tech_Index	14	0	31.74	4.26	15.93	10.30	22.02	29.85	34.75	65.00	54.70

Variable	IQR
Tech_Index	12.73

4.3.2.2 Step 2 – Fit a Second-Order Model to Estimate the Effects and Determine the Significant Ones

Before analyzing the extreme vertices design, we must specify the mixture components within Minitab.

 To define the mixture design, go to:

Stat > DOE > Mixture > Define Custom Mixture Design

Under **Components**, select the factors A, B, and C. If there are any process variables (Section 4.3.1.3, step 3), you have to specify them under **Process variables**. If you have a mixture-amount experiment (Section 4.3.1.4, step 4), under **Mixture Amount** select **In Column** and specify the column including the different mixture amounts. In our design we don't have process variables and the mixture amount is constant. Therefore, leave the field under **Process Variables** blank and under **Mixture Amount**, check the option **Constant**. Click **Lower/Upper**. For each component in the list, verify that the high and low limit settings are correct. If you have linear constraints (Section 4.3.1.7, step 7) select **Linear Constraints** and specify them. In our design we don't have linear constraints. Click **OK**.

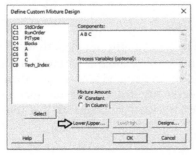 

In the main dialog box, click on **Designs**. Specify the columns corresponding to the standard order, run order, and point type of the runs. Click **OK** in each dialog box. Now you are ready to analyze the mixture design.

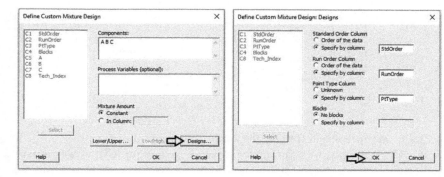

To analyze the mixture design, go to:

Stat > DOE > Mixture > Analyze Mixture Design

Select the response variable "Tech_Index" under **Responses**. Under **Type of Model**, select **Mixture components only**. With process variables or different mixture amounts check the corresponding option. Under **Analyze Components in**, choose **Pseudocomponents**, then click on **Terms**. At the top of the next dialog box, in **Include component terms for model,** specify **Quadratic**. Click **OK** and in the main dialog box, choose **Graphs**. Under **Residual Plots**, choose **Four in one**. Click **OK** in each dialog box and Minitab shows the results of the analysis both in the Session window and through the required graphs.

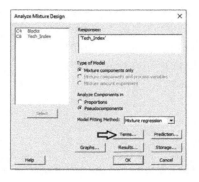
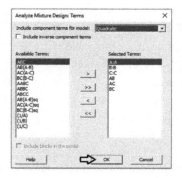

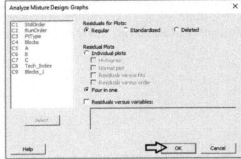

4.3.2.2.1 Interpret the Results of Step 2

In the ANOVA table, we examine the p-values to determine whether the associations between any terms in the model and the response are statistically significant. Remember that the p-value is a probability that measures the evidence against the null hypothesis. Lower probabilities provide stronger evidence against the null hypothesis. The null hypothesis is that there is no association between the term and the response. Usually we consider a significance level $\alpha = 0.05$, but in an exploratory phase of the analysis we may also consider a significance level equal to 0.10. When the p-value is greater than or equal to alpha, we fail to reject the null hypothesis. When it is less than alpha, we reject the null hypothesis and claim statistical significance.

Regression for Mixtures: Tech_Index versus A; B; C

Analysis of variance for Tech_Index (pseudocomponents)

Source	DF	Seq SS	Adj SS	Adj MS	F-Value	p-Value
Regression	5	3083.77	3083.8	616.75	22.85	0.000
Linear	2	2383.49	1572.2	786.09	29.12	0.000
Quadratic	3	700.28	700.3	233.43	8.65	0.007
A*B	1	0.20	163.7	163.68	6.06	0.039
A*C	1	146.25	163.2	163.24	6.05	0.039
B*C	1	553.83	553.8	553.83	20.52	0.002
Residual error	8	215.94	215.9	26.99		
Lack-of-fit	4	111.67	111.7	27.92	1.07	0.474
Pure error	4	104.27	104.3	26.07		
Total	13	3299.71				

The table does not display p-values for the linear terms of the components in mixture experiments because of the dependence between the components.

Additionally, the model does not include a constant because it is incorporated into the linear terms.

If an interaction term that includes only mixture components shows a p-value less than the significance level, the association between the blend of components and the response is statistically significant.

If your model includes process variables, significant interaction terms that include components and the process variables indicate that the effect of the components on the response variable depends on the process variables.

Looking at the ANOVA table, all the interaction terms between mixture components show a p-value less than 0.05. This means that the mean technical performance index for the two-blend mixture is statistically different from the value you would obtain by calculating the simple mean of the response variable for each pure mixture.

Note that in the ANOVA table, Minitab displays the *lack-of-fit test* when your data contain replicates. This test helps investigators to determine whether the model accurately fits the data. Its null hypothesis states that the model accurately fits the data. As usual, compare the p-value to your significant level α. If the p-value is less than α, reject the null hypothesis and conclude that the lack of fit is statistically significant. If the p-value is larger than α, you fail to reject the null hypothesis and you can conclude that the lack of fit is not statistically significant.

In our example, the p-value (0.474) is greater than 0.05: we conclude that there is no strong evidence of lack of fit. Lack-of-fit can occur if important terms such as interactions are not included in the model or if the fitted model shows several unusually large residuals (Stat Tool 2.9).

In addition to the results of the analysis of variance, Minitab displays some other useful information in the Model Summary table.

The quantity **R-squared** (R-sq, R^2) is interpreted as the percentage of the variability among technical performance data, explained by the terms included in the ANOVA model.

Model summary

S	R-sq	R-sq(adj)	PRESS	R-sq(pred)
5.19542	93.46%	89.37%	925.580	71.95%

The value of R^2 varies from 0 to 100%, with larger values being more desirable.

The adjusted R^2 (R-sq(adj)) is a variation of the ordinary R^2 that is adjusted for the number of terms in the model. Use adjusted R^2 for more complex experiments with several factors, when you want to compare several models with different numbers of terms.

The value of **S** is a measure of the variability of the errors that we make when we use the ANOVA model to estimate the performance data. Generally, the smaller it is, the better the fit of the model to the data.

Before proceeding with the results reported in the Session window, take a look at the residual plots. A **residual** represents an **error** that is the distance between an observed value of the response and its estimated value by the ANOVA model. The graphical analysis of residuals based on **residual plots** helps you discover possible violations of the ANOVA underlying assumptions (Stat Tools 2.8, 2.9).

A check of the normality assumption could be made looking at the histogram of the residuals (lower left), but with small samples, often the histogram shows irregular shape.

The normal probability plot of the residuals may be more useful. Here we can see a tendency of the plot to follow the straight line.

The other two graphs (Residuals versus Fits and Residuals versus Order) seem unstructured, thus supporting the validity of the ANOVA assumptions. In the Residual versus Fits graph, we notice the two ouliers.

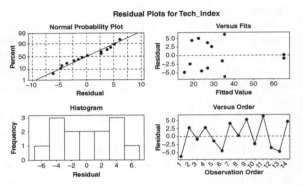

Residual Plots for Tech_Index

Estimated regression coefficients for Tech_Index (pseudocomponents)

Term	Coef	SE Coef	t-Value	p-Value	VIF
A	26.74	3.60	*	*	1.68
B	37.49	9.09	*	*	5.38
C	64.95	3.59	*	*	1.86
A*B	−66.6	27.1	−2.46	0.039	3.25
A*C	−41.6	16.9	−2.46	0.039	1.65
B*C	−109.9	24.3	−4.53	0.002	3.81

Returning to the Session window, we find two tables of coefficients expressed as pseudocomponents and as proportions. Look at the sign of the coefficients of the interactions of two components. Negative coefficients indicate that the two components act against each other. That is, the mean response value for the two-component blend is less than the value obtained by calculating the simple mean of the response variable for each pure mixture.

Estimated regression coefficients for Tech_Index (component proportions)

Term	Coef
A	463.82
B	1038.56
C	457.53
A*B	−1665.42
A*C	−1040.08
B*C	−2748.19

Positive coefficients indicate that the two components act in synergy. That is, the mean response value for the two-component blend is greater than the value you would obtain by calculating the simple mean of the response variable for each pure mixture.

4.3.2.3   Step 3 – Optimize the Response

After fitting the model, additional analyses can be carried out to find the optimum set of mixture components that optimizes the response (Stat Tool 3.8).

To optimize the response variable, go to:

Stat > DOE > Mixture > Response Optimizer

In the table, under **Goal**, select one of the following options for the response:

- **Do not optimize**: Do not include the response in the optimization process.
- **Minimize**: Lower values of the response are preferable.
- **Target**: The response is optimal when values meet a specific target value.
- **Maximize**: Higher values of the response are preferable.

In our case study the goal is to maximize the technical performance: choose **Maximize** from the drop-down list and click **Setup**. In the next window you need to specify a response target and lower or upper bounds depending on your goal (Stat Tool 3.8).

- If your goal is to maximize the response (larger is better), you need to specify the lower bound and a target value. You may want to set the lower value to the smallest acceptable value and the target value at the *point of diminishing returns*, i.e. a point at which going above its value does not make much difference. If there is no point of diminishing returns, use a very high target value.
- If your goal is to target a specific value, you should set the lower value and upper value as points of diminishing returns.
- If the goal is to minimize (smaller is better), you need to specify the upper bound and a target value. You may want to set the upper value to the largest acceptable value and the target value at the point of diminishing returns, i.e. a point at which going below its value does not make much difference. If there is no point of diminishing returns, use a very small target value.

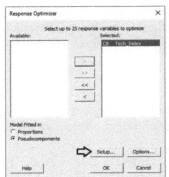
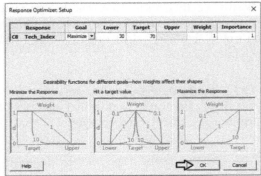

In our example, we can set Lower at 30 and the Target at 70.

Note that in the process of response optimization, Minitab calculates and uses desirability indicators from 0 to 1 to represent how well a combination of factor levels satisfies the goals you have for the responses. Specifying a different *weight* varying from 0.1 to 10, you can emphasize or deemphasize the importance of attaining the goals. Furthermore, when you have more than one response, Minitab calculates individual desirabilities for each response and a composite desirability for the entire set of responses. With multiple responses, you may assign a different level of *importance* to each response by specifying how much effect each response has on the composite desirability.

In our example, leave **Weight** and **Importance** equal to **1**. Click **OK** in each dialog box. In the optimization plot, Minitab will display the optimal solution, i.e. the component setting which maximizes the composite desirability.

The optimization plot shows how different component settings affect the response. The vertical lines on the graph represent the current component settings. The numbers displayed at the top of a column show the current mixture (in red). The horizontal blue lines and numbers represent the responses for the current mixture. The optimal solution is plotted on the graph and serves as the initial point. The settings can then be modified interactively, to determine how different mixtures affect the response, by moving the red vertical lines corresponding to each component. To return to the initial settings, right-click and select **Navigation** and then **Reset to Initial Settings**.

4.3.2.3.1 Interpret the Results of Step 3

The optimization plot is a useful tool to explore the sensitivity of the response variable to changes in the component settings and in the neighborhood of a local solution, for example the optimal one.

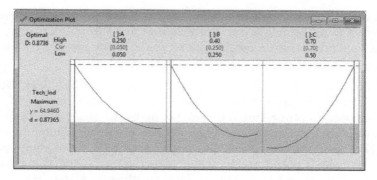

Response Optimization

Parameters

	Goal	Lower	Target	Upper	Weight	Import
Tech_Index	Maximum	30	70	70	1	1

Global Solution

Components

A	=	0.05
B	=	0.25
C	=	0.7

Predicted Responses

Tech_Index = 64.9460 , desirability = 0.873650

Composite Desirability = 0.873650

The Session window displays:

- information about the boundaries, weight, and importance for the response variable (Parameters);
- the solution that maximizes the response (Global Solution);
- the fitted response for the global solution and the value of the desirability function.

For technical performance data, the global solution has component A set at 0.05, component B at 0.25 and C at 0.70 (components are expressed as proportions). For this mixture the fitted response (mean technical performance index) is equal to 64.9. The composite desirability (0.87) is quite close to 1 denoting a satisfactory overall achievement of the specified goal.

4.3.2.4 Step 4 – Examine the Shape of the Response Surface and Locate the Optimum

Graphic techniques such as *contour, surface,* and *overlaid plots* help us to examine the shape of the response surface in regions of interest. Through these plots, you can explore the potential relationship between *pairs of factors* and the responses. You can display a contour, a surface or an overlaid plot considering two factors at a time, while setting the remaining factors to predefined levels. With contour and surface plots, you consider a single response; in overlaid plots, you can consider more than one response.

To display the contour and surface plots, go to: **Stat > DOE > Mixture > Contour/Surface Plots**

Choose **Contour plot** and click **Setup**. In the dialog box, select the response "Tech_Index" beside **Response**. Under **Components or Process Variables**, check the option **Select a triplet of components for a single plot**. If you have more than three components in the mixture, you can select the option **Generate plots for all triplets of components**. Under **Component Unit in Plot(s)**, check **Proportion**. Click **Contours**, and under **Data Display**, choose **Contour lines** and **Symbols at design points**. Click **OK** in the last two dialog boxes. In the main dialog box select **Surface Plot** and click **Setup**.

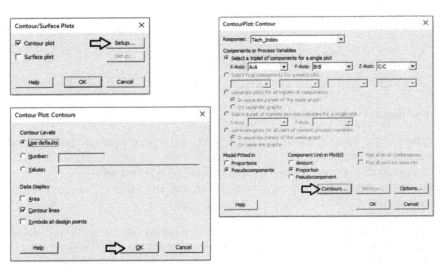

In the dialog box, select the response "Tech_Index" in **Response**. Under **Components or Process Variables**, check the option **Select a triplet of components for a single plot**. If you have more than three components in the mixture, you can select the option **Generate plots for all triplets of components**. Under **Component Unit in Plot(s)**, check **Proportion**. Click **OK** in each dialog box.

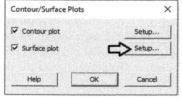
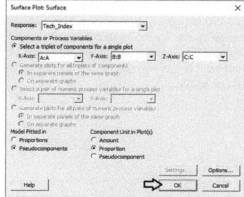

To change the appearance of the surface plot, double-click on any point inside a surface.

4.3.2.4.1 Interpret the Results of Step 4

In the contour plot, the technical performance index is represented by different level curves. Double-click on any point in the frame and select **Crosshairs** to interactively view variations in the response. In the surface plot the performance index is represented by a smooth surface. Both plots show how the component proportions are related to the performance. To maximize performance, researchers should choose proportions for the components in the lower right corner of the design space where the performance index is highest. It is relatively easy to see that the optimum is very close to 8 for additive A and to 7 for additive B, and to explore how the technical performance changes while increasing or decreasing factor levels.

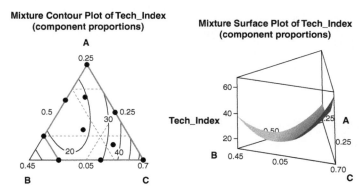

In the contour plot the solid gray contour represents the design space for this mixture design. However, you can see that the plot extends beyond the design space. In the surface plot the boundaries of the design space are not shown. Generally, you should use proportions that are in the design space because the relationships between the variables are uncertain outside the design space.

5

Product Validation

5.1   Introduction

Product validation is a key phase of product development. It is when the success criteria of the developed product are tested and evaluated by consumers to ensure expectations and product potential are explored before launching onto the market. This phase requires a combination of different tests to prove product performance involving both technical tests to establish key variables and tests with consumers to ensure it is actually perceived. In general it is key to assess: (i) performance and efficacy of the product against criteria connected to consumer claims and regulation requirements, (ii) consumer reaction to the product, and (iii) stability over time to ensure performance can be maintained. In all three of these key aspects, it is important to understand how response variables relate to each other and that predictions for variables are within the relevant performance range. This is possible using linear or multiple linear regression models.

Three case studies are discussed in this chapter. The first concerns a multi-center randomized pilot study performed to assess the effect of a new pharmaceutical preparation as add-on therapy in GERD (gastroesophageal reflux disease) patients. Based on data retrieved from a questionnaire designed to measure and evaluate specific GERD symptoms of heartburn, regurgitation, and dyspepsia, three scores were calculated by combining frequency and severity of each symptom. Investigators need to analyze the relationship between these GERD scores and establish whether the score for heartburn symptoms can be explained by the other two scores. The example shows how linear regression analyses can help to solve such problems.

The second and third case studies refer to the application of linear regression models in Stability Studies, used to analyze the stability of a product over time and to determine the product's shelf life (i.e. the length of time that a response is expected to remain within desired specifications). A typical stability study

End-to-End Data Analytics for Product Development: A Practical Guide for Fast Consumer Goods Companies, Chemical Industry and Processing Tools Manufacturers, First Edition.
Rosa Arboretti, Mattia De Dominicis, Chris Jones, and Luigi Salmaso.
© 2020 John Wiley & Sons Ltd. Published 2020 by John Wiley & Sons Ltd.
Companion website: www.wiley.com/go/salmaso/data-analytics-for-pd

tests multiple batches of a product over time to determine the shelf life. Through a linear model, investigators can represent the relationship between the response variable, the time variable, and an optional batch factor. This batch factor can be fixed if investigators sample all of the batches available for the product under study, or random if the batches are a random sample from a larger population of possible batches. As an example of a fixed batch factor, we consider a quality engineer who wants to determine the shelf life of lozenges that contain a new component, taking into account that the component's concentration in the lozenges (mg/loz) decreases over time. Considering four pilot batches available for production, the engineer wants to determine if and for how long the component concentration remains between 23.4 and 28.6 mg/loz. The engineer wants to test one lozenge from each batch at six different times at three-month intervals. He or she needs first to create a stability study data collection worksheet and then to analyze the collected data to estimate the new component's shelf life.

As an example of random batch factor, we consider a pharmaceutical company that wants to explore if the viscosity of a drug is shown to be stable and between 8000 and 40 000 mPa.s throughout the first six months under accelerated conditions. Considering a sample of three batches from those available for production, the viscosity of a 75 ml quantity of drug is measured at five different times at varying intervals. The example shows first how to create a stability study data collection worksheet and then how to analyze the collected data to estimate the drug's shelf life.

In short, the chapter deals with the following:

Topics	Stat tools
Correlation and regression analysis, correlation coefficient	5.1, 5.3
Scatterplot	5.2
Regression models, simple linear regression models	5.4, 5.5
Goodness of fit, residual analysis	5.6, 5.7
Multiple linear regression models	5.8

Learning Objectives and Outcomes

Upon completion of this chapter, you should be able to do the following:

Identify relationships between variables using correlation and regression techniques.

Use correlation as an indication of the strength and direction of a linear relationship between two variables.

Fit regression models to study how much a dependent variable varies based on changes to one or more independent variables and predict outcomes.

Evaluate the assumptions of a regression analysis.

Collect and analyze stability data for the estimation of products' shelf life.

5.2 Case Study: GERD Project

A multicenter randomized pilot study was performed to assess the effect of a new pharmaceutical preparation as add-on therapy in GERD patients with inadequate response to daily proton pump inhibitor treatment. At baseline, during a seven-day run-in period, 141 patients completed a specific symptom questionnaire designed to measure and evaluate specific GERD symptoms of heartburn, regurgitation, and dyspepsia. Three scores were calculated, one for each symptom, by combining the frequency and severity of each. Investigators need to analyze the *relationship* between the GERD scores. In particular, they wish to establish whether the score related to heartburn symptoms can be explained by the other two scores.

To solve this problem, we can apply a regression analysis in which the variable Heartburn is the response and the other two scores, Regurgitation and Dyspepsia, are the explanatory variables (Stat Tool 5.1).

The variables' setting is the following:

- Variables Heartburn, Regurgitation, and Dyspepsia are *continuous quantitative variables.*

Column	Variable	Type of data	Label
C1	SUBJECT_ID		Patients' code
C2	Heartburn	Numeric data	Score based on heartburn symptoms
C3	Regurgitation	Numeric data	Score based on regurgitation symptoms
C4	Dyspepsia	Numeric data	Score based on dyspepsia symptoms

File: Gerd_Project.xlsx.

5.2.1 Evaluation of the Relationship among Quantitative Variables

To study the relationship among the three GERD symptom scores, let's proceed in the following way:

Step 1 – Perform an exploratory analysis through scatterplots (Stat Tool 5.2) and calculate the correlation coefficients (Stat Tool 5.3).
Step 2 – Build a multiple linear regression model (Stat Tools 5.4–5.8).
Step 3 – If required, reduce the model to include the significant terms.
Step 4 – Predict response values.
Step 5 – Explore the response surface in multiple linear regression.

5.2.1.1 Step 1 – Perform an Exploratory Analysis through Scatterplots and Calculate the Correlation Coefficients

To detect a possible association between two *quantitative* variables, the first thing to do is to plot the data on a *scatterplot* (Stat Tool 5.2). This helps us to explore the direction, strength, and form of any potential relationship. When a linear relationship is a plausible model to represent the trend of data, the correlation coefficient (Stat Tool 5.3) helps us to quantify this possible linear relationship.

To display the scatterplots, go to:

Graph > Scatterplot

Check the graphical option **Simple.** In the table in the next dialog box, you can choose up to 20 scatterplots, specifying the X and Y variables for each of them. Under **Y variables**, select Heartburn for the first and second scatterplots, and under **X variables**, select Regurgitation for the first and Dyspepsia for the second. Click **Multiple Graphs**, and in the next screen choose the option **In separate panels of the same graph**. Then click **OK** in each dialog box.

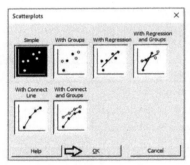

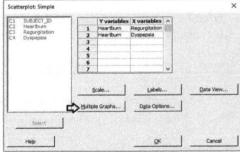

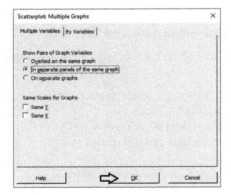

To calculate the correlation coefficients, go to: **Stat** > **Basic Statistics** > **Correlation**

Under **Variables**, select Heartburn, Regurgitation, and Dyspepsia. As they are quantitative continuous variables, choose **Pearson correlation** in **Method**. Check the option **Display p-values** and click **OK**.

5.2.1.1.1 Interpret the Results of Step 1 The scores related to the symptoms of both regurgitation and dyspepsia seem to be positively related to those of heartburn, since the points form an upward pattern from left to right (Stat Tool 5.2). However, the points are scattered, indicating variability in the data.

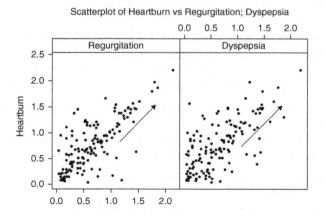

Correlations

	Heartburn	Regurgitation
Regurgitation	0.755	
	0.000	
Dyspepsia	0.688	0.605
	0.000	0.000

Cell contents

Pearson correlation

p-value

The correlation coefficient (Pearson correlation) between the Heartburn and Regurgitation scores is equal to 0.76, which indicates that the two variables have a reasonably strong linear relationship (Stat Tool 5.3). Setting the significance level α at 0.05, the p-value (0.000) is less than 0.05. The linear relationship between the two variables is statistically significant.

The correlation coefficient between Heartburn and Dyspepsia is equal to 0.69, which indicates that the two variables have a slightly weaker statistically significant relationship (p-value less than 0.05).

Stat Tool 5.1 Correlation and Regression Analysis

We frequently encounter variables that seem to be related:

E.g. Satisfaction scores may be linked to the age of customers.

Blood pressure may be affected by temperature.

The evaluation of relationships among variables may be useful in:

- An exploratory analysis, when we know very little about the variables, to describe possible relationships among them.
- At a most advanced phase of a statistical analysis, to detect aspects that can explain a variable of interest and make a prediction about its values.

Regression and *correlation analysis* help us best evaluate relationships among variables.

When we believe one or more variables may influence other variables, helping us to explain or predict them, we can apply a *regression analysis*:

- The variables we wish to explain or understand are called *dependent* or *response variables*.
- The variables that help to explain or predict the response variables are called *independent predictors* or *explanatory variables*.

Stat Tool 5.1 (Continued)

➤ *Example 5.1*. A research team needs to investigate how age, gender, and having children may influence consumer satisfaction with a new disinfectant spray for home hygiene.

In this example, age, gender, and having children are independent variables and satisfaction is the response.

When interest lies in the evaluation of relationships among variables without specifying their roles in terms of dependent and independent variables, a *correlation analysis* may be performed.

➤ *Example 5.2*. Research aims to analyze how sports activities and eating habits in preschool children and their parents are mutually related.

In this context, investigators are not interested in identifying a response that is explained by other variables, but simply to study whether being active is accompanied by a healthy diet.

Stat Tool 5.2 Scatterplot

To detect a possible association between *two quantitative variables*, the first graphical tool we can use is the *scatterplot*. By using Cartesian coordinates, each value of one variable is displayed on the horizontal axis and the corresponding value of the other variable is plotted on the vertical axis.

➤ *Example 5.3*. Thermoplastic elastomer (TPE) sheets can perform optimally in terms of energy return and lightness for professional and amateur athletes. Consider six tests during which a research team measures the thickness (mm) of TPE sheets used in insoles and their cushioning energy for walking (CEW, in mJ) (see Table 5.1).

Table 5.1 Raw data.

Test	Thickness (mm)	CEW (mJ)
1	6.97	122.76
2	7.02	123.29
3	7.05	126.98
4	6.93	113.87
5	7.03	129.74
6	6.97	112.34

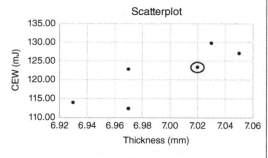

Figure 5.1 Scatterplot of thickness vs. CEW; each point representing one observation.

Stat Tool 5.2 (Continued)

Each point in the scatterplot (Figure 5.1) represents one observation: e.g. test number 2 is at 7.02 mm on the horizontal axis and 123.29 on the vertical axis.

A scatterplot is the first *descriptive* tool we use to evaluate a possible relationship between two quantitative variables and shows patterns in the relationship that would not be seen by just looking at the data. This exploratory tool enables us to see patterns in the data and helps guide further statistical analyses. A scatterplot gives information about three characteristics of a relationship:

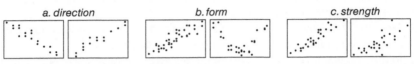

a. direction   b. form   c. strength

a. The *direction of the relationship* is the easiest characteristic to spot on a scatterplot and may be:

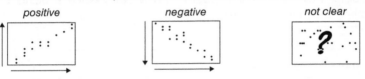

positive   negative   not clear

The points are plotted from the lower-left corner to the upper-right corner. The variables move in the same direction. As one variable increases, the other one increases.	The points are plotted from the upper-left corner down to the lower-right corner. The variables move in the opposite direction. As one variable increases, the other one decreases.	No clear relationship exists between variables: the points fall randomly on the graph, with no pattern in either direction.

b. The *form of the relationship* is the next characteristic to evaluate, where form is the pattern the data forms. The form of a relationship may be:

linear

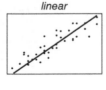

The points follow a straight line, suggesting a linear relationship between the variables.

The points don't follow a straight line, e.g. the pattern follows an upside-down U-shaped curve.

non linear

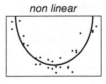

Notice that when the points follow a straight line but the line is *flat*, it means that there is no relationship between the two variables: as one variable increases, the other one does not vary and remains substantially constant. A linear relationship may be:

Stat Tool 5.2 (Continued)

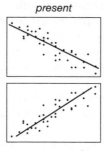

The points follow a nonflat, straight line with a positive or negative slope. There is a linear relationship.

The points follow a flat straight line with a slope equal to 0. There is no relationship.

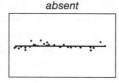

c. The third characteristic to consider is the *strength of the relationship*. Consider a linear relationship. To evaluate strength, consider how closely the data points follow a straight line. The strength of a relationship may be:

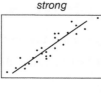

The two variables are closely related. The relationship is strong because the points closely follow a straight-line.

The variables are still related. However, the points are scattered around the line, indicating variability in the data and, therefore, a weak relationship between variables.

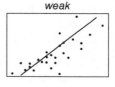

➤ *Example 5.3.* Consider again the example related to the thickness (mm) of TPE sheets used in insoles.

The variables are moving in the same direction. As the thickness increases, so does CEW; as CEW decreases, so does thickness. The scatterplot shows a *positive* relationship between thickness and CEW.

The data seems to follow a straight line, suggesting a *linear* relationship between the variables.

The points follow a linear pattern and are located quite close

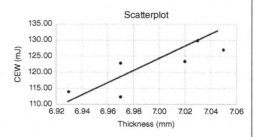

Figure 5.2 Scatterplot of thickness vs. CEW showing a moderate linear relationship.

to that line. The scatterplot shows a *moderate* linear relationship between thickness and CEW (Figure 5.2).

You will learn more on how to *measure* the strength and direction of a linear relationship in the next session.

Stat Tool 5.3 Correlation Coefficient

When a scatterplot shows a *linear relationship* between two quantitative variables, we can measure the direction and strength of the association by calculating the *correlation coefficient* (*also known as Pearson correlation*).

The correlation coefficient, symbolized by *r* for sample data and by the Greek letter ρ (rho) for population, quantifies the linear association between two quantitative variables.

The correlation coefficient only measures linear relationship. Although other important nonlinear relationships may exist, we cannot use correlation to investigate them.

The correlation coefficient is a single value ranging from −1 to +1.

The *sign* of the coefficient indicates the *direction* of the relationship:

- If one variable tends to increase as the other decreases, the coefficient is *negative*.
- If one variable tends to increase as the other increases, the coefficient is *positive*.
- If there is no linear relationship between variables, the coefficient equals 0.

The *magnitude* of the correlation coefficient indicates the *strength* of the association (Figure 5.3):

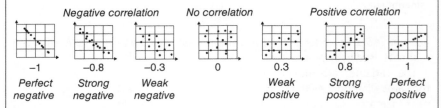

Figure 5.3 Strength and direction of correlations.

- When there is a perfect linear relationship, all points fall on a straight line, and the correlation equals either +1 or −1.
- For a *stronger* relationship, the correlation is closer to +1 or −1.
- A correlation near zero indicates a weak linear relationship.

> *Example 5.3.* Let's return to the example related to the thickness (mm) of TPE sheets used in insoles.

The points follow a *linear pattern* and are located quite close to that line (Figure 5.4).

Stat Tool 5.3 (Continued)

The *correlation coefficient* is equal to 0.834, which indicates that the two variables have a *strong* relationship in a *positive* direction.

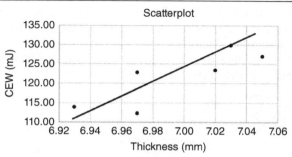

Figure 5.4 Scatterplot of thickness vs. CEW showing a strong, positive relationship.

Bear in mind that determining how strong the correlation is between variables is not only a statistical matter but also depends on the application. In some applications, 0.834 may be considered a weak correlation. Consider a laboratory testing the calibration of new equipment to standard measures. Here the technicians are hoping for a correlation of at least 0.998 and would consider anything less to be too weak to certify the new equipment.

After calculating the correlation coefficient by using *sample data,* we can apply a *hypothesis test* (Stat Tools 1.15 and 1.16) to determine whether a linear relationship between the variables exists. For this test, the null and the alternative hypotheses are the following:

Null hypothesis	Alternative hypothesis
No linear correlation between the variables exists.	A linear correlation between the variables exists.
$H_0: \rho = 0$	$H_1: \rho \neq 0$

Setting the significance level α at 0.05 or 0.01:

- If the p-value is less than α, reject the null hypothesis and conclude that the linear correlation between the two variables is statistically significant.

  p-value $< \alpha$ Reject the null hypothesis that the population correlation coefficient equals 0.

- If the p-value is greater than or equal to α, fail to reject the null hypothesis and conclude that the linear correlation between the two variables is NOT statistically significant.

  p-value $\geq \alpha$ Fail to reject the null hypothesis that the population correlation coefficient equals 0.

➤ *Example 5.3.* Let's return to the example related to the thickness (mm) of TPE sheets used in insoles.

Stat Tool 5.3 (Continued)

Suppose the research team repeats the experiment, this time measuring the thickness (mm) of TPE sheets used in insoles and their cushioning energy for walking (mJ) in 20 trials. Calculate the correlation coefficient:

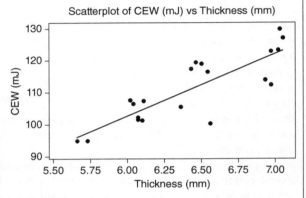

Correlations

Pearson correlation   0.841

p-value   0.000

The correlation coefficient is equal to 0.841, which indicates that the two variables have a strong relationship in a positive direction. Setting the significance level α at 0.05, the p-value is lower. We can reject the null hypothesis. The linear correlation between the two variables is statistically significant (see Figure 5.5).

Figure 5.5 Scatterplot of thickness vs. CEW showing a very strong relationship.

5.2.1.2   Step 2 – Build a Multiple Linear Regression Model

When we believe one or more variables may influence other variables, we can use *regression models* (Stat Tool 5.4) to better understand this kind of relationship.

Furthermore, if the data form a linear pattern, we can use *linear regression models* (Stat Tools 5.5–5.8). In our case study we need to establish whether the score related to the Heartburn symptoms (response) can be explained by the scores related to the Regurgitation and Dyspepsia symptoms (explanatory variables). From the exploratory analysis, a linear relationship can be a reasonable model.

 To build a linear regression model, go to:

Stat > Regression > Regression > Fit Regression Model

Select the numeric response variable "Heartburn" under **Responses**. Under **Continuous predictors**, select "Regurgitation" and "Dyspepsia" and click on **Model**. In the next window, highlight the two explanatory variables "Regurgitation" and "Dyspepsia" and next to **Interactions**

through order: 2, select **Add** to include the main effects and the two-way interaction between the two predictors in the model. Click **OK** in each dialog box.

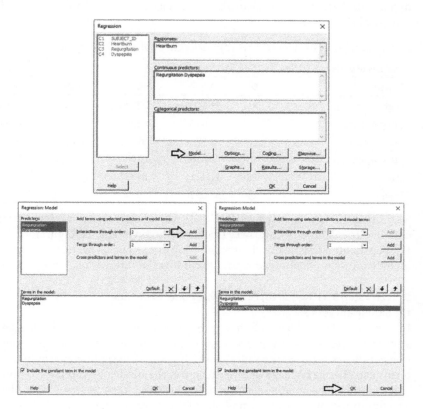

5.2.1.2.1 *Interpret the Results of Step 2* In the analysis of variance (ANOVA) table we examine the p-values to determine whether any predictor or interaction is statistically significant. Remember that the p-value is a probability that measures the evidence against the null hypothesis. Lower probabilities provide stronger evidence against the null hypothesis.

For the main effects, the null hypothesis is that there is no linear relationship between a predictor and the response.

For the two-way interaction, H_0 states that the relationship between a predictor and the response does not depend on the other predictor in the term.

We usually consider a significance level α equal to 0.05, but in an exploratory phase of the analysis we may also consider a significance level α of 0.10.

When the p-value is greater than or equal to alpha, we fail to reject the null hypothesis. When it is less than alpha, we reject the null hypothesis and claim statistical significance.

Regression analysis: Heartburn versus Regurgitation; Dyspepsia.

Analysis of variance

Source	DF	Adj SS	Adj MS	F-value	p-value
Regression	3	21.6704	7.22346	86.64	0.000
Regurgitation	1	3.0337	3.03369	36.39	0.000
Dyspepsia	1	1.3771	1.37708	16.52	0.000
Regurgitation*Dyspepsia	1	0.0105	0.01046	0.13	0.724
Error	137	11.4216	0.08337		
Total	140	33.0920			

So, setting the significance level α to 0.05, which terms in the model are significant in our example? The answer is: only the main effects of Regurgitation and Dyspepsia are statistically significant. You may want to reduce the model to include only significant terms, thus excluding the interaction term.

5.2.1.3   Step 3 – If Required, Reduce the Model to Include the Significant Terms

In our example, we have only one term to remove from the model, but generally with more terms to remove, it is advisable not to remove entire groups of nonsignificant terms at the same time. The statistical significance of individual terms can change because of other terms in the model. To reduce your model, you can use an automatic selection procedure, the *stepwise strategy*, to identify a useful subset of terms, choosing one of the three commonly used alternatives (standard stepwise, forward selection, and backward elimination).

 To reduce the model, go to:

Stat > Regression > Regression > Fit Regression Model

Select the numeric response variable "Heartburn" under **Response**. Under **Continuous predictors**, select "Regurgitation" and "Dyspepsia" and choose **Stepwise**. In **Method** select **Backward elimination** and in **Alpha to remove** specify **0.05**, then click **OK**. In the main dialog box, choose **Graphs**. Under **Residual plots** choose **Four in one**. Minitab will display several residual plots to examine whether your model meets the assumptions of the analysis (Stat Tool 5.7). Click **OK** in each dialog box.

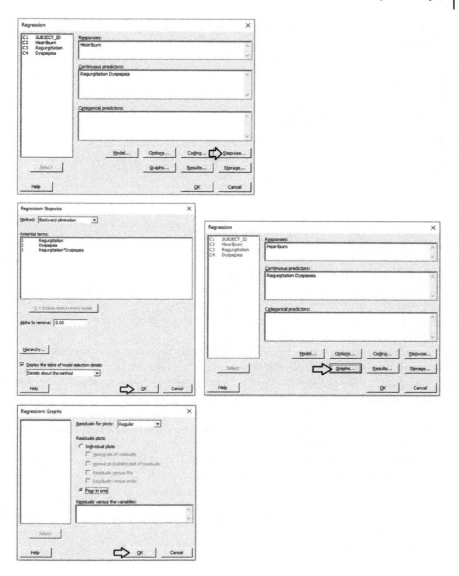

Complete the analysis by adding the factorial plots that show the relationships between the response and the significant terms in the model, displaying how the response changes as the predictors change.

To display factorial plots, go to:

Stat > Regression > Regression > Factorial Plots

Move the variables "Regurgitation" and "Dyspepsia" from **Available:** to **Selected:** using the button **>>** and then click **OK**.

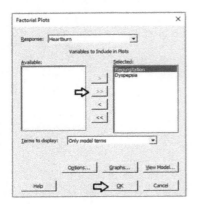

5.2.1.3.1 Interpret the Results of Step 3
Setting the significance level alpha to 0.05, the ANOVA table shows the significant terms in the model. Remember that the stepwise procedure may add nonsignificant terms in order to create a hierarchical model. In a hierarchical model, all lower-order terms that comprise a higher-order term also appear in the model.

In addition to the results of the ANOVA, Minitab displays some other useful information in the Model Summary table to evaluate its goodness of fit (Stat Tool 5.6).

The quantity **R-squared** (R-sq, R^2) is interpreted as the percentage of the variability among Heartburn scores, explained by the terms included in the model.

The value of R^2 varies from 0% to 100%, with larger values being more desirable. The model that includes Regurgitation and Dyspepsia explains 65.45% of the variation of Heartburn.

The adjusted R^2 (R-sq(adj)) is a variation of the ordinary R^2 that is adjusted for the number of terms in the model. Use adjusted R^2 when you want to compare several models with different numbers of terms.

Regression Analysis: Heartburn versus Regurgitation; Dyspepsia
Backward Elimination of Terms
α to remove = 0.05
ANOVA

Source	DF	Adj SS	Adj MS	F-Value	p-Value
Regression	2	21.660	10.8300	130.73	0.000
Regurgitation	1	6.006	6.0056	72.50	0.000
Dyspepsia	1	2.776	2.7764	33.52	0.000
Error	138	11.432	0.0828		
Total	140	33.092			

Model Summary

S	R-sq	R-sq(adj)	R-sq(pred)
0.287822	65.45%	64.95%	63.54%

The value of **S** is a measure of the variability of the errors that we make when we use the linear model to estimate the Heartburn scores. Generally, the smaller it is, the better the fit of the model to the data.

Before proceeding with the results reported in the Session window, take a look at the residual plots. A *residual* represents an *error*, i.e. the distance between an observed value of the response and its value estimated by the model. The graphical analysis of residuals based on *residual plots* helps discover possible violations of the underlying assumptions (Stat Tool 5.7).

A check of the normality assumption could be made by looking at the histogram of the residuals (lower left), but with small samples, often the histogram shows irregular shape.

The normal probability plot of the residuals may be more useful. Here we can see a tendency of the plot to follow

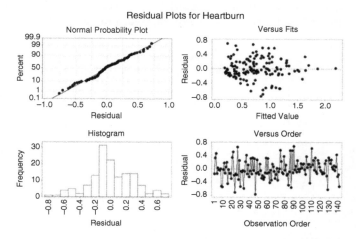

the straight line. The assumption of normality can reasonably be considered satisfied.

The other two graphs (Residuals versus Fits and Residuals versus Order) seem unstructured, thus supporting the validity of the linear regression assumptions.

In particular, the plot of residuals versus the fitted values doesn't display any recognizable pattern. The constant variance assumption is not violated. Some points are quite distant from the other points and may be outliers. The plot of Residuals versus Order shows a random pattern of residuals on both sides of 0. The independency assumption is not violated.

Returning to the Session window, we find the **regression equation** derived from the data that expresses the relationship between the response and the important predictors.

In the regression equation, we have a coefficient for each term in the model. They are expressed in the original measurement units. These coefficients are also displayed in the Coefficients table. The intercept (0.1317) is the value of Heartburn when the other two scores are equal to 0. Recall that sometimes the intercept has no practical meaning, but gives the line a better position across data points. The coefficient of Regurgitation (+0.5837) is positive. This means that as Regurgitation scores increase, the mean Heartburn score increases (positive relationship). The magnitude of the slope (0.5837) means that, keeping the value of Dyspepsia constant, as Regurgitation scores increase by 1 point, the mean Heartburn score increases by 0.5837. The coefficient of Dyspepsia (+0.3731) is positive. This means that as Dyspepsia scores increase, the mean Heartburn score increases (positive relationship). The magnitude of the slope (0.3731) means that, keeping the value of Regurgitation constant, as Dyspepsia scores increase by 1 point, the mean Heartburn score increases by 0.3731. Look also at the factorial plots (Main Effects Plots) to display how the mean Heartburn score linearly varies as predictors change.

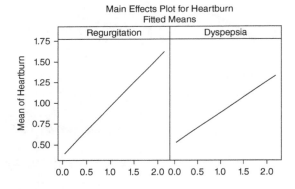

Coefficients

Term	Coef	SE coef	t-Value	p-Value	VIF
Constant	0.1317	0.0456	2.89	0.004	
Regurgitation	0.5837	0.0686	8.51	0.000	1.58
Dyspepsia	0.3731	0.0644	5.79	0.000	1.58

Regression Equation

Heartburn = 0.1317 + 0.5837 Regurgitation + 0.3731 Dyspepsia

Earlier we learned how to examine the p-values in the ANOVA table to determine whether an explanatory variable is statistically related to the response. In the Coefficients table, we find the same results in terms of tests on coefficients and we can evaluate the p-values to determine whether they are statistically significant. The null hypothesis is that there is no linear effect on the response. Setting the significance level α at 0.05, if a p-value is less than α, reject the null hypothesis and conclude that the linear effect on the response is statistically significant.

Both for Regurgitation and Dyspepsia the p-value (0.000) is less than 0.05. The linear relationships between the response variable and the two explanatory variables are statistically significant.

5.2.1.4 Step 4 – Predict Response Values

Use the regression equation to predict values (fitted values) of the response variable Heartburn for combinations of explanatory variables of interest. This fitted value is obtained by inserting the specified values of the explanatory variables into the regression equation.

 To predict response values, go to:

Stat > Regression > Regression > Predict

Suppose that you need to predict the value of Heartburn when Regurgitation and Dyspepsia are both equal to 0.40. Under Regurgitation and Dyspepsia. specify the value 0.4 and click **OK**.

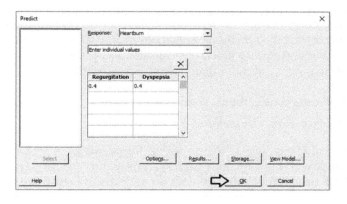

5.2.1.4.1 *Interpret the Results of Step 4* Suppose that you need to predict the mean response when Regurgitation and Dyspepsia are both equal to 0.40. We obtain a predicted value of 0.51 for the mean Heartburn score.

You can also see two confidence intervals (Stat Tool 1.14) for the predicted value. The confidence interval (CI) of the prediction represents a range within which the mean response is likely to fall given specified settings of the predictors. The prediction interval (PI) represents a range within which a single new observation is likely to fall given specified settings of the predictors.

Prediction for Heartburn

Regression Equation

Heartburn = 0.1317 + 0.5837 Regurgitation + 0.3731 Dyspepsia

Settings

Variable	Setting
Regurgitation	0.4
Dyspepsia	0.4

Prediction

Fit	SE fit	95% CI	95% PI
0.514461	0.0284568	(0.458194; 0.570729)	(−0.0574243; 1.08635)

5.2.1.5   Step 5 – Explore the Response Surface in Multiple Linear Regression

In multiple linear regression with more than one predictor, graphic techniques such as *contour, surface*, and *overlaid plots* help us to examine the response surface in regions of interest. Through these plots, you can explore the potential relationship between *pairs of explanatory variables* and the response. You can display a contour, a surface, or an overlaid plot considering two predictors at a time, while setting the remaining explanatory variables to predefined values. With contour and surface plots, you consider a single response; in overlaid plots, you may consider more than one response together.

 To display the contour plots, go to:

Stat > Regression > Regression > Contour Plot

In the dialog box, select "Heartburn" in **Response**. Here we have only two predictors, but with more than two explanatory variables, you have to check the option **Generate plots for all pairs of continuous variables** to create a contour plot for each pair of predictors. Click **Contours**, and under **Data Display**, choose **Contour lines**. Click **OK**. With only two predictors, click **OK** in the main dialog box. Or, if you have more than two predictors, select **Settings** in the main dialog box and set the remaining explanatory variables at specified values. Click **OK** in each dialog box.

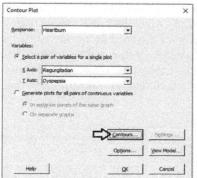

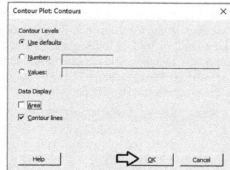

To display the surface plots, go to:

Stat > Regression > Regression > Surface Plot

In the dialog box, select "Heartburn" in **Response**, and click **OK**. With more than two predictors check the option **Generate plots for all pairs of continuous variables**. Then select **Settings** to set the other explanatory variables at specified values and click **OK**.

To change the appearance of the surface plots, double-click on any point inside the surface. In **Attributes**, under **Surface Type**, select the option **Wireframe** instead of the default **Surface**. Then click **OK.**

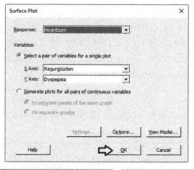

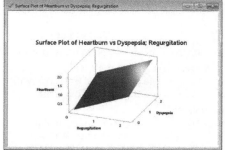

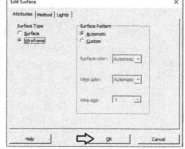

5.2.1.5.1 Interpret the Results of Step 5 Contour and surface plots are useful tools to explore how a response variable changes while the predictors increase or decrease. In the Contour plot, Regurgitation and Dyspepsia scores are plotted on the x- and y-axes and the mean Heartburn score is represented by different level contours. Double-click on any point in the frame and select **Crosshairs** to interactively view variations in the response. In surface plots, the mean Heartburn score is represented by a hyperplane.

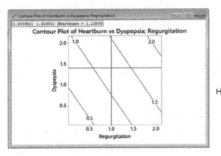
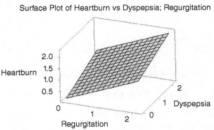

Stat Tool 5.4 Regression Models

So far, we have learned to graph the relationship between two numerical variables using a scatterplot and to measure the strength and direction of that relationship with a correlation coefficient.

What more can we learn about the relationship among our variables?

When we believe one or more variables may influence other variables, i.e. there are variables that help us to explain or predict other variables, we can use a third tool: a *regression model*. Regression models can help us better understand the relationship among variables.

In regression models:

- Variables we wish to explain or understand are called *dependent* or *response variables*.
- Variables that help to explain or predict the response variables are called *independent predictors* or *explanatory variables*.

A regression model can be:

Simple	Multiple	Multivariate
With 1 response variable (e.g. growth of leaves) and 1 predictor (e.g. amount of sunlight)	With 1 response variable (e.g. growth of leaves) and more than 1 predictor (e.g. amount of sunlight, amount of water)	With more than 1 response variable (e.g. growth of leaves, growth of roots) and more than 1 predictor (e.g. amount of sunlight, amount of water)

Stat Tool 5.4 (Continued)

If a *linear* relationship may be a reasonable pattern for our data, we can use *linear regression models.*

Linear regression provides an equation or model to describe the relationship among quantitative variables.

A linear regression model can be:

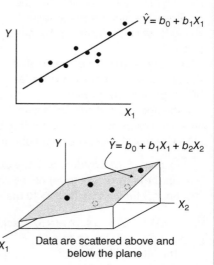

$$\hat{Y} = b_0 + b_1 X_1$$

- *Simple,* providing an equation corresponding to the line that best represents the relationship between the response (Y) and only one explanatory variable (X).
- *Multiple,* with more than one predictor. For example, a multiple linear regression with two explanatory variables calculates an equation corresponding to the plane that best represents the relationship between the response (Y) and the two explanatory variables (X_1, X_2).

$$\hat{Y} = b_0 + b_1 X_1 + b_2 X_2$$

Data are scattered above and below the plane

Stat Tool 5.5 Simple Linear Regression Models

Simple linear regression calculates an equation corresponding to the line that best represents the relationship between the response (*Y*) and the explanatory variable (X_1).

b_0 is the *intercept*, which is the value of *Y* when X_1 is zero. Sometimes the intercept does not have much practical use.

b_1 is called the *slope* or *regression coefficient*. The slope is perhaps the most important aspect of the equation because it indicates the impact of the explanatory variable on the response. The slope is the amount *Y* changes when X_1 increases by one unit.

The steeper the slope, the greater the effect of the explanatory variable on the response variable.

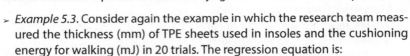

If the line is flat, the slope is zero, which indicates no linear relationship between the variables.

Notice that the response variable always goes on the vertical axis (*y*-axis).

> *Example 5.3*. Consider again the example in which the research team measured the thickness (mm) of TPE sheets used in insoles and the cushioning energy for walking (mJ) in 20 trials. The regression equation is:

$$\text{CEW (mJ)} = -14.67 + 19.57 \,\text{Thickness (mm)}$$

The intercept (−14.67) is the value of CEW (mJ) when thickness is 0. In this context, intercept has no practical meaning, but gives the line a better position across data points (Figure 5.6). The slope (+19.57) is positive. This means that as thickness increases, CEW

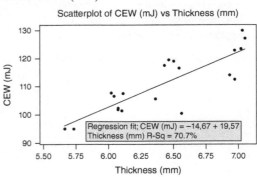

Figure 5.6 Scatterplot of thickness vs. CEW with regression line.

Stat Tool 5.5 (Continued)

increases (positive relationship). The magnitude of the slope (19.57) means that as thickness increases by 1 mm, CEW increases by 19.57 mJ.

We can also use regression to make *predictions* about the response based on values of the explanatory variables.

Suppose that the research team is interested in predicting the value of CEW when the thickness is equal to 6.25 mm (note that this value was not observed in the experiment).

To make this prediction, we plug 6.25 into the equation:

$$CEW\,(mJ) = -14.67 + 19.57 \times 6.25 = 107.64\,mJ$$

We can find the same result by looking at the regression line drawn through the scatterplot (Figure 5.7).

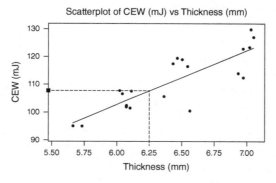

Figure 5.7 Scatterplot of thickness vs. CEW with regression line, CEW = 6.25 mm.

After finding a regression line that best represents the data, how do we know if the regression line is statistically significant? To determine the statistical significance of the equation, we can conduct a *hypothesis test* (Stat Tools 1.15 and 1.16) on the slope of the regression equation.

The slope b_1 of the regression line calculated by using sample data is a point estimate of the unknown population slope β_1. We will test whether the population slope is equal to 0. Remember that if the line is flat, the slope is zero and no linear relationship exists between the variables. But if the line is not flat, the slope is not zero, and a relationship may exist between the variables.

After calculating the slope by using sample data, we can apply a hypothesis test to determine whether a linear relationship between the response and the predictor exists. For this test, the null and the alternative hypotheses are the following:

Stat Tool 5.5 (Continued)

Null hypothesis	Alternative hypothesis
No linear relationship exists between the variables.	A linear relationship exists between the variables.
$H_0: \beta_1 = 0$	$H_1: \beta_1 \neq 0$

Setting the significance level α usually at 0.05 or 0.01:

- If the p-value is less than α, reject the null hypothesis and conclude that the linear relationship between the two variables is statistically significant.

  p-value $< \alpha$ — Reject the null hypothesis that the population slope equals 0.

- If the p-value is greater than or equal to α, fail to reject the null hypothesis and conclude that the linear relationship between the two variables is NOT statistically significant.

  p-value $\geq \alpha$ — Fail to reject the null hypothesis that the population slope equals 0.

> *Example 5.3.* Consider again the example in which the research team measured the thickness (mm) of TPE sheets used in insoles and the cushioning energy for walking (mJ) in 20 trials.

The Minitab output for the test on the coefficients of the regression equation (Figure 5.8) is the following:

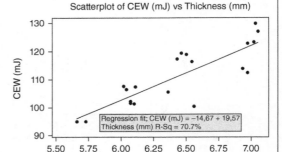

Figure 5.8 Scatterplot of thickness vs. CEW showing a statistically significant linear relationship.

Coefficients

Term	Coef	SE coef	t-value	p-value
Constant	−14.7	19.1	−0.77	0.453
Thickness (mm)	19.57	2.97	6.60	0.000

Consider the results of the hypothesis test on the slope. Let's focus on the p-value. Setting the significance level α at 0.05, the p-value (0.000) is lower. We can reject the null hypothesis. There is a statistically significant linear relationship between the thickness of TPE sheets used in insoles and the cushioning energy for walking.

Stat Tool 5.6 Goodness of Fit	

Once we know that the relationship between the variables is statistically significant, we can determine how well the response is explained by the explanatory variable.

In other words, we need to evaluate the *goodness of fit* of the regression model.

With any regression analysis, some points are closer to the line and some are farther. The closer the points are to the line, the better the regression line fits the data.

The distances between the points (observed data from the sample) and the regression line are called *residuals*. They represent the portion of the response that is not explained by the regression equation.

Residuals enable us to check the equation to see if the line shows a good fit to the data.

> *Example 5.3.* Consider again the example in which six tests were performed by a research team to study the association between the thickness (mm) of TPE sheets used in insoles and the cushioning energy for walking (mJ) (see Table 5.2).

Table 5.2 Raw data.

Test	Thickness (mm)	CEW (mJ)
1	6.97	122.76
2	7.02	123.29
3	7.05	126.98
4	6.93	113.87
5	7.03	129.74
6	6.97	112.34

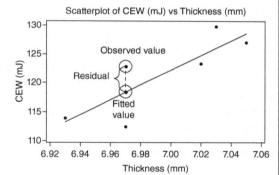

Figure 5.9 Residuals.

Observe the first row. Thickness is 6.97 and CEW is 122.7, as shown on the plot (Figure 5.9). This data point is called the *observed value* because it is the response value obtained from the sample.

The regression equation is: CEW (mJ) = −776.1 + 128.3 Thickness (mm)

What if we plug this observed value into the equation for the regression line? The value the equation returns is called the *fitted value*: CEW = −776.1 + 128.3 × 6.97 = 118.15.

We can see that for Thickness = 6.97 the regression equation predicted a value for CEW of 118.15, but the observed value of CEW was 122.76. The difference (122.76 − 118.15 = 4.61) between the observed and the fitted value is the *residual*.

To measure how much variability in the response is explained by the exploratory variable, we can use a measure called *coefficient of determination*, symbolized as R^2.

R^2 is a value between zero and one and is usually expressed as a percentage to make it easier to interpret. As a percentage, R^2 is a value between 0% and 100%.

Stat Tool 5.6 (Continued)

100%
↑
If the explanatory variable explains much of the variation of the response, the points on the plot are close to the regression line: the residuals are small and R^2 has a high value (high goodness of fit).

R^2
↓
0%
If the explanatory variable only explains a small amount of the variation in the response, the points are farther from the regression line: the residuals are larger and R^2 has a low value (low goodness of fit).

➤ *Example 5.3.* For the previous example, R^2 is equal to 70.73%. The thickness explains 70.73% of the variation of CEW, the points on the plot are fairly close to the regression line, the residuals are small, and R^2 has a fairly high value (good fit).

Stat Tool 5.7 Residual Analysis

In the previous sections, we fitted a regression line to represent the relationship between two quantitative variables. We performed a hypothesis test to determine the line's significance and calculated R^2 to evaluate the goodness of fit.

Does regression analysis require any assumptions about data, and if so, how can we verify them?

With regression, the following *three assumptions* about *errors* can potentially be made when using a regression model to estimate the response variable:

1) Errors are random and independent.
2) Errors are normally distributed.
3) Errors have constant variance across all values of X.

To check the assumptions, we evaluate the error in our regression model by examining the *residuals*. Remember that the residuals represent the difference between the observed and fitted values.

The easiest way to evaluate the residuals is by using *residual plots* (*residual analysis*).

1) The first assumption specifies that errors are *random* and *independent*. A prerequisite for the first assumption is that *random samples* have been collected (Stat Tool 1.2).
   During regression analysis, we can also plot the *residuals in the order that the data were collected* and look for any patterns. If data were not collected in any meaningful order, or if the order the data were collected is unknown, do not use this plot.

Stat Tool 5.7 (Continued)

Structureless plots, showing no obvious patterns, reveal no serious problems of lack of independence. Look at the two *residuals versus order plots* below (Figure 5.10). On the left, the points on the plot are distributed in a random order and do not exhibit any patterns. The residuals appear to be *independent* of one another. The plot on the right, however, displays a recognizable *pattern*: from the 11th observation onwards, the points gradually increase over time, thus indicating that the residuals may not be independent.

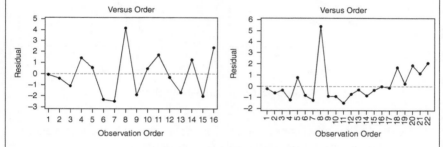

Figure 5.10 Residuals versus order plots.

If we detect patterns in the residuals versus order plots, we should determine the cause. Possible causes are problems during data collection.

2) The second assumption states that the errors follow a *normal* distribution. The *normal distribution* is a bell-shaped and symmetric distribution widely used in statistics to represent quantitative variables showing symmetry and unimodal distributions (Figure 5.11).

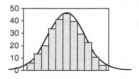

Figure 5.11 Normal distribution.

A check of the *normality assumption* could be made through a *normal probability plot* of the residuals. If the underlying response distribution is normal, this plot will resemble a straight line. If the residuals approximately follow the *straight line on the plot*, you don't have a significant violation of the normality assumption.

Stat Tool 5.7 (Continued)

Below are two normal probability plots (Figure 5.12). The one on the left represents residuals that *likely follow a normal distribution*, where residuals form approximately a straight line. The one on the right exhibits a pattern that indicates the residuals *do not follow a normal distribution*.

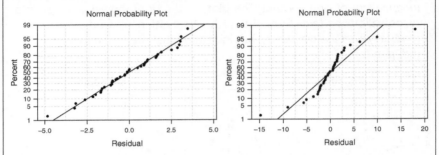

Figure 5.12 Normal probability plots.

3) The third assumption states that the *variance or spread* in the errors across the values of the explanatory variable X_1 is *constant*.

To test this assumption, we can plot the *residuals versus the fitted values* (Figure 5.13). Notice that the residual values are on the vertical axis and the fitted values are on the horizontal axis. Remember that the fitted values are obtained using the regression equation.

Below are two *residuals versus the fitted values* plots. The one on the left demonstrates constant variance across the X values. The scatter of the points around the line is similar across all values of the predictor, and no unusual patterns in the data are present.

The one on the right shows a common problem: the residuals increase systematically with the fitted values, which is sometimes called the funnel or megaphone effect. The variance is not constant across the X values.

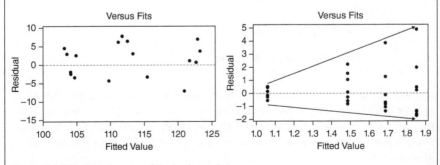

Figure 5.13 Residuals versus fitted values plots.

Stat Tool 5.8 Multiple Linear Regression Models

In simple linear regression, we have only one explanatory variable.

Sometimes we may be interested in two or more explanatory variables. In those cases we should use *multiple linear regression models*.

The steps to follow are the same as those we presented earlier:

1) Fit the regression model by estimating one coefficient (slope) for each explanatory variable.
2) Check the statistical significance of each coefficient (slope) by hypothesis testing.
3) Check the goodness of fit of the model.
4) Check the assumptions of linear regression models by performing a residual analysis.

5.3 Case Study: Shelf Life Project (Fixed Batch Factor)

Linear regression models (Stat Tools 5.4–5.8) have important applications in stability studies, used to analyze the stability of a product over time and to determine the product's shelf life, i.e. the length of time that a response is expected to remain within desired specifications. A typical stability study tests multiple batches of a product over time to determine the shelf life (e.g. investigators test two pills from each of four batches every two weeks). Through a linear model, investigators can represent the relationship between the response variable, the time variable, and an optional batch factor. Stability studies usually refer to a *quantitative continuous* response variable. The batch factor can be *fixed* or *random*: it is fixed if investigators sample all of the batches available for the product under study; it is random if the batches are a random sample from a larger population of possible batches.

A quality engineer wants to determine the shelf life of lozenges that contain a new component. The concentration of the component in the lozenges decreases over time. Considering four pilot batches available for the production, the engineer wants to determine if and how long the component concentration remains within 23.4 and 28.6 mg/loz. The engineer wants to test one lozenge from each batch at six different times at three-month intervals. First they need to create a stability study data collection worksheet and then analyze the collected data to estimate the shelf life of the new component.

To study the shelf life of the new component, let's proceed in the following way:

Step 1 – Create a data collection worksheet.
Step 2 – Apply a stability analysis to estimate the shelf life.
Step 3 – Predict response values.

5.3.1.1   Step 1 – Create a Data Collection Worksheet

Let's begin by creating a data collection worksheet, considering four batches, six different times at three-month intervals and only one replicate (one measure) for each trial.

To create the data collection worksheet, go to:

Stat > Regression > Stability Study
 > Create Stability Study Worksheet

In the drop-down menu at the top, choose **Test times in numeric format, at constant intervals.** You can choose different options according to your specific interests. In **Number of test times**, select **6**, then specify **0** and **3** as **Starting time** and **Interval length**. In **Unit of time** specify **Month**. Set **4** as **Number of batches** and in **Number of samples from each batch at each time**, choose **1**. Click **Options**. In the next window select the option **Randomize batch sequence and repeats at each testing time** and click **OK** in each dialog box.

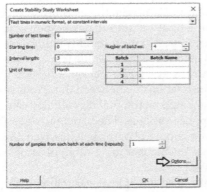
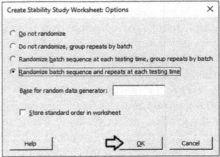

Minitab shows the data collection form in the worksheet (only the first 10 runs are shown here as an example). You can then proceed by collecting the data for the response variable following the order provided by the column "RunOrder" and entering the values in the column "Response".

↓	C1	C2	C3-T	C4	C5
	RunOrder	Month	Batch	Response	
1	1	0	1	*	
2	2	0	3	*	
3	3	0	4	*	
4	4	0	2	*	
5	5	3	4	*	
6	6	3	1	*	
7	7	3	2	*	
8	8	3	3	*	
9	9	6	4	*	
10	10	6	3	*	

5.3.1.2   Step 2 – Apply a Stability Analysis to Estimate the Shelf Life

To estimate the shelf life of the new component and determine if and how long the component concentration remains within 23.4 and 28.6 mg/loz, you can apply a stability analysis with a fixed batch factor. We fit a linear model to represent the relationship between the response variable (component concentration), the time variable, and the batch factor.

The variables setting is the following:

- Variable "Month" is a *discrete quantitative variable*, expressed in months.
- Variable "Response" is a *continuous quantitative variable*, expressed in mg/loz.

Column	Variable	Type of data	Label
C1	Runorder		Order of trial execution
C2	Month	Numeric data	Time of response measurement in months
C3-T	Batch	Categorical data	Number of batch
C4	Response	Numeric data	Component concentration in mg/loz

File: Shelf_Life_fixed_Project.xlsx.

 To apply the stability analysis with a fixed batch factor, go to:

Stat > Regression > Stability Study > Stability Study

Select the response variable "Response", the time variable "Month", the batch factor "Batch" and specify the lower and upper specification limits. Click **Options**. In the drop-down menu at the top, choose **Batch is a fixed factor**. Under **Percent of response within spec limits with specified confidence**, select **50%** to calculate the shelf life based on the mean of the response values. Take into account that Minitab considers a symmetric normal distribution (Stat Tools 1.5 and 5.7) so that the 50th percentile, which corresponds to the median is equal to the mean

(Stat Tool 1.6). In **Confidence level**, enter the level of confidence for the confidence interval or leave the default value 95%. In **Alpha for pooling batches**, enter the level of significance for model selection or leave the default value 0.25. During model selection, if the p-value for each term in the model is greater than or equal to the alpha level you specify, then the term is removed from the model. Then click **OK**. In the main window, click **Graphs**. Under **Shelf life plot**, in the second drop-down list, select **No graphs for individual batches**, and under **Residual Plots**, select **Four in one**. Click **OK** in each dialog box.

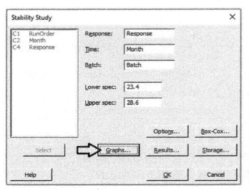

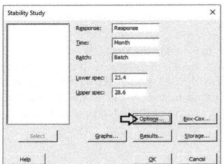
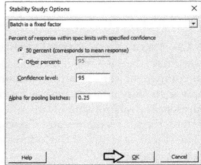

5.3.1.2.1 *Interpret the Results of Step 2*

The Model Selection table shows the results to determine whether the association between the response and the terms in the model (time, batch, and their interaction) is statistically significant. Take into account that during the model selection process, Minitab considers a significance level alpha equal to 0.25 (generally used in stability analysis) to include terms in the model and creates a hierarchical model. In a hierarchical model, all lower-order terms that comprise a higher-order term, are included in the model: you can see that the p-value for the Month by Batch interaction is 0.000, so both "Month" and "Batch" are in the model (they would have been included even if one of them had been singularly not significant). To further examine the relationship with the response, also consider the Regression Equation table.

Model Selection with $\alpha = 0.25$

Source	DF	Seq SS	Seq MS	F-value	p-value
Month	1	31.0889	31.0889	510.15	0.000
Batch	3	7.0812	2.3604	38.73	0.000
Month*Batch	3	2.4644	0.8215	13.48	0.000
Error	16	0.9750	0.0609		
Total	23	41.6096			

Regression Equation

Batch	
1	Response = 26.548 − 0.2152 Month
2	Response = 26.629 − 0.1838 Month
3	Response = 26.343 − 0.3257 Month
4	Response = 25.429 − 0.1638 Month

When the batch by time interaction term is included in the final model, Minitab displays a separate equation for each batch where batches have different intercepts and different slopes. If the batch factor is included in the final model, but the batch by time interaction is not included, then all batches have different intercepts but the same slope. Lastly, if time is the only term in the model, then all batches share the same intercept and slope, and Minitab displays a single regression equation.

In our example, the regression equations for each batch have different intercepts and slopes. Note that as expected, all the slopes are negative, thus indicating that the component concentration reduces with time. Batch 3 has the steepest slope, −0.3257, which indicates that, every three months, the component concentration for Batch 3 decreases by 0.3257 mg/loz. Batch 4 has the smallest intercept, 25.429 mg/loz, which indicates that Batch 4 had the lowest concentration at time zero.

Minitab uses the final model to estimate shelf life for each batch. The Shelf Life Estimation table shows the specification limits and the shelf life estimates. If the batch factor is not included in the final model, then the shelf life is the same for all batches. Otherwise, the shelf life for each batch is different and the overall shelf life is equal to the smallest of the individual shelf life values.

In our example, Batch 3 has the shortest shelf life estimate equal to 8.34 months, so the overall shelf life is estimated as 8.34 months.

Shelf Life Estimation

Batch	Shelf life
1	13.147
2	15.424
3	8.3698
4	10.828
Overall	8.3698

Lower spec limit = 23.4, upper spec limit = 28.6, shelf life = time period in which you can be 95% confident that at least 50% of response is within spec limits

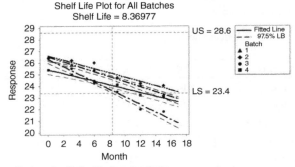

LS = Lower Specification, US = Upper Specification

To determine how well the model fits your data, examine the goodness-of-fit statistics (Stat Tool 5.6) in the Model Summary table. You see that both R^2 and adjusted R^2 are close to 100, which indicates that the model fits the data well. Remember also that the value of **S** is a measure of the variability of the errors that we make when we use the linear model to estimate the component concentrations. Generally, the smaller it is, the better the fit of the model to the data.

Model Summary

S	R-sq	R-sq(adj)	R-sq(pred)
0.246861	97.66%	96.63%	94.49%

To discover possible violations of the underlying assumptions of the regression model used to estimate the shelf life (Stat Tool 5.7), let's have a look at the residual plots.

A check of the normality assumption could be made looking at the histogram of the residuals (bottom left), but with small samples, often the histogram has an irregular shape.

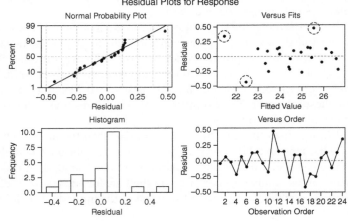

In the normal probability plot, we can see a tendency of the plot to bend slightly upward on the right, but the plot is not in any case grossly non-normal.

The other two graphs (Residuals versus Fits and Residuals versus Order) seem unstructured, thus supporting the validity of the linear regression assumptions.

In the Residuals versus Fits plot, the only noteworthy observation is the presence of three points (enclosed in dashed circles) that could be outliers. In these cases investigators may try to identify the cause of potential outliers.

5.3.1.3   Step 3 – Predict Response Values
Use the final model to predict values (fitted values) of the variable "Response" at a specified time of interest.

To predict response values, go to:

Stat > Regression > Stability Study > Predict

Since the batch variable is included in the final model, you can also predict the response for specific batches. Suppose that you need to predict the value of "Response" after 15 months for the first batch. Under **Month** specify the value **15** and under **Batch** enter the number of the batch. Click **OK**.

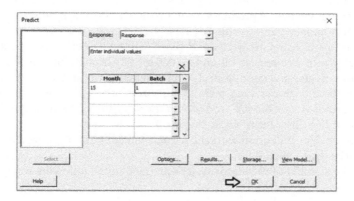

5.3.1.3.1   Interpret the Results of Step 3   The predicted value for the mean component concentration at month 15 for the first batch is equal to 23.319 mg/loz. You can also see two confidence intervals (Stat Tool 1.14) for the predicted value. The confidence interval of the prediction (95% CI) represents a range within which the mean response is likely to fall at that specified time. The prediction interval (95% PI) represents a range within which a single new observation is likely to fall at that time.

Prediction for Response

Regression Equation

Batch	
1	Response = 26.548 – 0.21524 Month

Settings

Variable	Setting
Month	15
Batch	1

Prediction

Fit	SE fit	95% CI	95% PI
23.3190	0.178665	(22.9403; 23.6978)	(22.6730; 23.9651)

5.4 Case Study: Shelf Life Project (Random Batch Factor)

The previous section shows how linear regression models (Stat Tools 5.4–5.8) are used in stability studies to examine the stability of a product over time and estimate the product's shelf life. The present case study considers a *random* batch factor: the batches of products used for the study are a random sample from a larger population of possible batches.

A pharmaceutical company wants to explore whether the viscosity of a drug remains stable and between 8000 and 40 000 mPa·s throughout the first six months under accelerated conditions. If confirmed, this would commonly lead to two-year stability under long-term conditions. Taking a sample of three batches from those available for production, the viscosity of 75 ml of the drug is measured at five different times at varying intervals. First we need to create a stability study data collection worksheet and then analyze the collected data to estimate the drug's shelf life.

To study the stability of the viscosity, let's proceed in the following way:

Step 1 – Create a data collection worksheet.
Step 2 – Apply a stability analysis to estimate the shelf life.

5.4.1.1 Step 1 – Create a Data Collection Worksheet
Let's begin by creating a data collection worksheet, considering three batches, five different times at varying intervals and only one replicate (one measure) for each trial.

To create the data collection worksheet, go to:

Stat > Regression > Stability Study
  > Create Stability Study Worksheet

In the drop-down menu at the top, choose **Test times in numeric format, at varying intervals**. You can choose different options according to your specific interests. In **Number of test times** and in **Unit of time** specify **Week**. Under **Week**, set the five time intervals at 0, 4, 12, 26, and 52 weeks. Select **3** as **Number of batches** and in **Number of samples from each batch at each time**, choose **1**. Click **Options**. In the next window select the option **Randomize batch sequence and repeats at each testing time** and click **OK** in each dialog box.

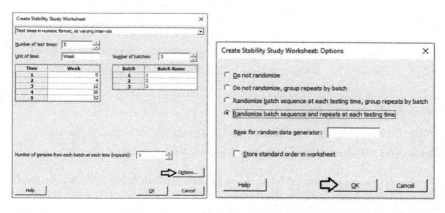

Minitab shows the data collection form in the worksheet (only the first 10 runs are shown here as an example). You can then proceed by collecting the data for the response variable following the order provided by the column "RunOrder" and entering the values in the "Response" column.

	C1	C2	C3-T	C4	C5
	RunOrder	Week	Batch	Response	
1	1	0	2	*	
2	2	0	1	*	
3	3	0	3	*	
4	4	4	2	*	
5	5	4	3	*	
6	6	4	1	*	
7	7	12	3	*	
8	8	12	1	*	
9	9	12	2	*	
10	10	26	1	*	

Worksheet 1 ***

5.4.1.2   Step 2 – Apply a Stability Analysis to Estimate the Shelf Life

To estimate the shelf life for the drug and determine if and how long the viscosity remains within 8000 and 40 000 mPa·s, you can apply a stability analysis with a random batch factor. We fit a linear model to represent the relationship between the response variable (viscosity), the time variable, and the batch factor.

The variables setting is the following:

- Variable "Week" is a *discrete quantitative variable*, expressed in weeks.
- Variable "Response" is a *continuous quantitative variable*, expressed in mPa·s.

Column	Variable	Type of data	Label
C1	Runorder		Order of trial execution
C2	Week	Numeric data	Time of response measurement in weeks
C3-T	Batch	Categorical data	Number of batch
C4	Response	Numeric data	Viscosity in mPa·s

File: Shelf_Life_random_Project.xlsx.

 To apply the stability analysis with a random batch factor, go to:

Stat > Regression > Stability Study > Stability Study

Select the response variable "Response", the time variable "Week", the batch factor "Batch" and specify the lower and upper specification limits. Click **Options**. In the drop-down menu at the top, choose **Batch is a random factor**. Under **Percent of response within spec limits with specified confidence**, by default Minitab uses the 95th percentile instead of the 50th percentile to calculate the shelf life. In **Confidence level**, enter the level of confidence for the confidence interval or leave the default value 95%. In **Alpha for pooling batches**, enter the level of significance for model selection or leave the default value 0.25. During model selection, if the p-value for each term in the model is greater than or equal to the alpha level you specify, then the term is removed from the model. Then click **OK**. In the main window, click **Graphs**. Here you can select *marginal* or *conditional* residuals to display on the residual plots, where marginal residuals are the difference between the fits and the observed values for the overall population, while conditional residuals are the difference between the fits and the observed values for the batches in the sample data. Use the conditional residuals to check the normality of the residuals. In **Residuals for plots**, choose **Conditional regular** and under **Residual Plots**, select **Four in one**. Click **OK** in each dialog box.

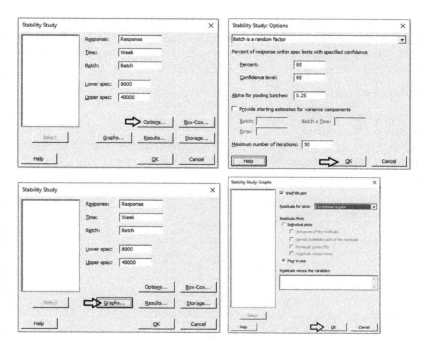

5.4.1.2.1 *Interpret the Results of Step 2*

The Model Selection table shows the model selection process during which Minitab considers a significance level alpha equal to 0.25 (generally used in stability analysis) to include terms in the model and create a hierarchical model (where all lower-order terms that comprise a higher-order term are included in the model). In our example, the batches and their interaction with time, don't show a significant impact on the response. A final model with pooled data and including only the time term is therefore estimated and Minitab specifies the result with the message: "*Terms in selected model: Week*".

Model Selection with α = 0.25

Model	−2 LogLikelihood	Difference	p-value
Week batch Week*batch	218.826		
Week batch	218.946	0.119770	0.836
Week	218.956	0.010098	0.460

Terms in selected model: Week

With a random batch factor, Minitab uses the final model to estimate only the *overall shelf life*, whether the batch factor is included or excluded from the model. The Shelf Life Estimation table and the Shelf Life Plot show the specification limits and the shelf life estimate. The shelf life, which is approximately 86 weeks, is an estimate of how long the investigators can be 95% confident that 95% of products are within the specification limits. This estimate applies to any batch randomly selected from the process.

Shelf Life Estimation

Lower spec limit = 8000
Upper spec limit = 40 000
Shelf life = time period in which you can be 95% confident that at least 95% of response is within spec limits
Shelf life for all batches = 86.3636

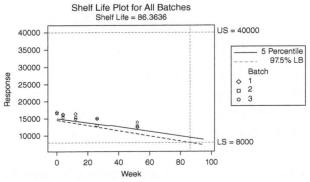

To determine how well the model fits your data, examine the goodness-of-fit statistics (Stat Tool 5.6) in the Model Summary table. You can see that both R^2 and adjusted R^2 are quite high, which indicates that the model fits the data quite well. Remember also that the value of **S** is a measure of the variability of the errors that we make when we use the linear model to estimate the viscosity. Generally, the smaller it is, the better the fit of the model to the data.

Model Summary

S	R-sq	R-sq(adj)	R-sq(pred)
712.184	77.68%	75.97%	70.35%

To discover possible violations of the underlying assumptions of the regression model used to estimate the shelf life (Stat Tool 5.7), let's have a look at the residual plots.

Residual Plots for Response

In the Normal Probability Plot, we can see a tendency of the plot to follow a straight line.

The other two graphs (Residuals versus Fits and Residuals versus Order) seem unstructured.

The linear regression assumptions can be considered satisfied.

If you want to estimate shelf life based on the 50th percentile instead of the 95th percentile, you can reapply the stability analysis without specifying the batch factor shown to be not significant in the previous analysis.

 To change the shelf life estimate percentile, go to:

Stat > Regression > Stability Study > Stability Study

Select the response variable "Response," the time variable "Week," and specify the lower and upper specification limits. Click **Options**. Under **Percent of response within spec limits with specified confidence**, select **50%** to calculate the shelf life based on the mean of the response values. Take into account that Minitab considered a symmetric normal distribution (Stat Tools 1.5 and 5.7) so that the 50th percentile which corresponds to the median is equal to the mean (Stat Tool 1.6). In **Confidence level**, enter the level of confidence for the confidence interval or leave the default value 95%. Click **OK** in each dialog box. In both the Session window and the Shelf Life Plot, Minitab shows the new shelf life estimate.

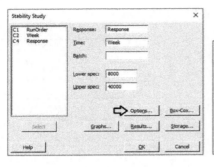
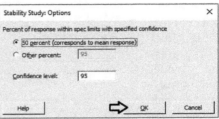

Shelf Life Estimation

Lower spec limit = 8000
Upper spec limit = 40 000
Shelf life = time period in which you can be 95% confident that at least 50% of response is within spec limits
Shelf life for all batches = 108.777

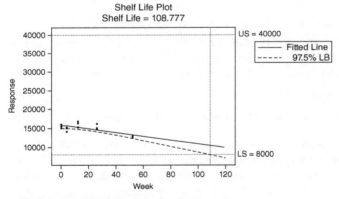

LS = Lower Specification, US = Upper Specification
Equation for fitted line: Response = 15771−48.9 Week

6

Consumer Voice

6.1 Introduction

Understanding how a new product, solution, or technology is delivering against consumer expectations is one of the key steps in innovation. Many different aspects can be considered, from consumer characteristics to product usage, features, and performance. The way in which they are dealt with in relation to expectation, preference, and purchase intent is key.

Previous products or a competitor's product can generally be benchmarked to assess the relative results and better predict market potential.

In many cases, product developers wish to evaluate the correlations between purchase intent, consumers, and product features or ways of using products. This is relevant to understanding what it is important to focus on in terms of consumer target (gender, age, economic situation, etc.), communication, and possible product improvements.

Very often, the consumer voice is assessed through categorical variables representing opinions, judgments, or preferences, measured on binary or ordinal scales, such as: satisfied/not satisfied; not preferred/slightly preferred/preferred/highly preferred.

In this chapter, several statistical techniques for categorical variables are reviewed to deal with different issues in the evaluation and optimization of consumer experience.

The first case study considers data from a consumer survey in which a sample of respondents were randomly selected to evaluate, by means of a questionnaire, their subjective assessments of the stain-removal performance of two detergents on washed fabrics. Among the various items in the questionnaire, consumers were asked to assign an overall satisfaction score from 1 (very dissatisfied) to 5 (very satisfied). Investigators want to establish whether the overall satisfaction score can be associated with the two products taking into

End-to-End Data Analytics for Product Development: A Practical Guide for Fast Consumer Goods Companies, Chemical Industry and Processing Tools Manufacturers, First Edition.
Rosa Arboretti, Mattia De Dominicis, Chris Jones, and Luigi Salmaso.
© 2020 John Wiley & Sons Ltd. Published 2020 by John Wiley & Sons Ltd.
Companion website: www.wiley.com/go/salmaso/data-analytics-for-pd

account different washing temperatures. In particular, they need to examine how the percentage of respondents with a satisfaction score of 4 or 5, denoted as "top two box," varies between the two products. Descriptive and graphical tools (cross tabulations, bar charts, and pie charts) and hypothesis testing (chi-square tests) can be very useful to explore relationships among categorical variables. Furthermore, logistic regression models can be applied for an in-depth evaluation of how satisfaction scores are related to different configurations of the product and to washing machine temperatures.

Again in relation to customer satisfaction evaluation, the final case study shows how the combination of design of experiments (DOE) techniques and logistic models can increase the opportunities for researchers to study the impact of each factor level on the response, while taking into account possible effects of other predictors. In the example, which considers two levels (absence/presence) for two additives, the researchers want to run a consumer test of the four different formulations of the laundry detergent, given by the combinations of the two additives. A random sample of consumers will test the four different formulations for a month, and at the end of the period, their overall opinion of the product will be recorded by means of a score from 1: dislike extremely, to 10: like extremely. The main purpose is to evaluate how different additive levels affect the probability of a liking score greater than or equal to 8, denoted as "top score."

In short, the chapter deals with the following:

Topics	Stat tools
Cross tabulations	6.1
Bar charts and pie charts	6.2
Chi-square test	6.3
Spearman rank correlation	6.4
Logistic regression models and odds ratios	6.5, 6.6

Learning Objectives and Outcomes

Upon completion of this chapter, you should be able to do the following:

Calculate and graphically display the frequency distribution of categorical variables.

Apply the chi-square test to evaluate the associations between two categorical variables.

Calculate and interpret the Spearman rank correlation as a measure of the association between ordinal variables.

Fit a logistic regression model to study the relationship between a binary response variable and one or more explanatory variables.

Interpret the meaning of odds ratios to quantify the relationship between a binary response variable and one explanatory variable.

6.2 Case Study: "Top-Two Box" Project

In a consumer survey a sample of 157 respondents were randomly selected to evaluate, by means of a questionnaire, their subjective assessments of stain-removal performance of two different detergents (A and B) on washed fabrics. Among the various items in the questionnaire, consumers were asked to assign an overall satisfaction score from 1 (very dissatisfied) to 5 (very satisfied). Investigators want to establish whether the overall satisfaction score can be associated with different washing temperatures. In particular, they need to examine how the percentage of respondents with a satisfaction score of 4 or 5, denoted as the "top two box," varies between the two products.

To solve this problem we can apply a logistic regression analysis where Satisfaction is the response and Product and Washing Machine Temperature (WM_temp) are the dependent variables (Stat Tool 6.5).

The variables setting is the following:

- Variable "Product" is a categorical variable, assuming two categories: A and B.
- Variable "WM_temp" is a quantitative discrete variable, assuming three levels: 30, 40, and 60 °C.
- Variable "Satisfaction" is a quantitative discrete variable, varying from 1 (very dissatisfied) to 5 (very satisfied).

Column	Variable	Type of data	Label
C1	ID_resp		Respondent's code
C2-T	Product	Categorical data	Product name
C3	WM_temp	Numeric data	Washing machine temperature in °C
C4	Satisfaction	Numeric data	Satisfaction score varying from 1 (very dissatisfied) to 5 (very satisfied)

File: Top_two_box_Project.xlsx

To study the relationship among the available variables, let's proceed in the following way:

Step 1 – Perform an exploratory descriptive analysis of the variable "Satisfaction" by product, through frequency tables, bar charts, and pie charts (Stat Tools 6.1 and 6.2).
Step 2 – Apply the chi-square (χ^2) test to evaluate the presence of association between the product and the satisfaction score (Stat Tool 6.3).
Step 3 – Perform an exploratory descriptive analysis of the variable "Satisfaction" by temperature and calculate the correlation coefficient between these two variables (Stat Tool 6.4).

Step 4 – Build a binary logistic regression model to examine how the product and the washing machine's temperature are related to the satisfaction score (Stat Tool 6.5).

6.2.1.1 Step 1 – Perform an Exploratory Descriptive Analysis of Satisfaction Scores by Product, Through Frequency Tables and Charts

In the Two-Top Box Project for the discrete variable "Satisfaction," use a bar chart or pie chart, stratifying by product to show the distribution of the satisfaction scores separately for each product. Add the frequency tables to complete the descriptive analysis.

To display the bar chart of "Satisfaction" by product, go to: **Graph > Bar Chart**.

Select the graphical option **Cluster** and click **OK**. In the next dialog box, in **Categorical variables**, choose the stratification variable "Product" and the response "Satisfaction." Click **Chart Options**.

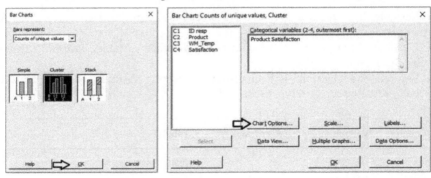

Under **Percent and Accumulate,** check the option **Show Y as Percent**, and under **Take Percent and/or Accumulate,** check **Within categories at level 1**. Click **OK** in each dialog box.

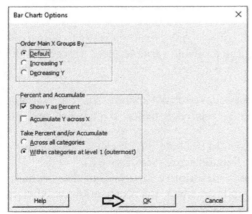

To display the frequency table of "Satisfaction" by product, go to: **Stat > Tables > Cross Tabulation and Chi-square.**

In **Rows**, select "Satisfaction" and in **Columns**, select "Product." Under **Display**, check **Counts** and **Column percents** and click **OK.**

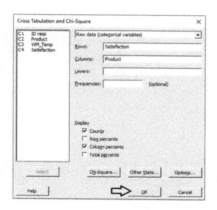

As the research team is particularly interested in the percentage of respondents belonging to the "top-two box" (i.e. customers assigning scores of 4 and 5 to their overall satisfaction), let's create a new variable "Top_two_Box" assigned the value "No" when "Satisfaction" is equal to 1, 2, or 3, and "Yes" when the satisfaction score is in the top-two box.

To create a new variable, go to: **Data > Recode > To Text**

Under **Recode values in the following columns**, select the variable "Satisfaction" and in **Method**, choose the option **Recode ranges of values.** Specify the lower and upper limit corresponding to the new categories "No" and "Yes." In **Endpoints to include**, choose **Both endpoints**, then under **Storage location for the recoded columns**, select **In specified columns of the current worksheet**, and finally under **Columns**, write the name of the new variable "Top_two_Box." Click **OK**. The new categorical variable will appear in column C5 of the worksheet.

To display the pie chart of "Top_two_Box" by product, go to: **Graph > Pie Chart**

In **Categorical variables**, choose the variable "Top_two_Box" and click **Labels**. At the top of the window, select **Slice Labels** and in the next dialog box check the option **Percent**. Click **OK** and in the main dialog box click **Multiple Graphs**. In **Show Pie Charts from Different Variables**, check the option **On the same graph**, then at the top of the window click **By Variables**. In the next window, under **By variables with groups on same graph**, select the stratification variable "Product" and click **OK** in each dialog box.

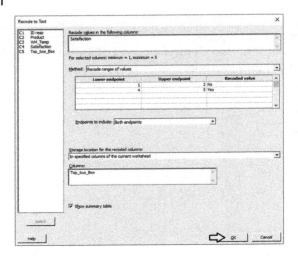

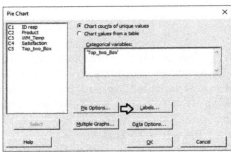

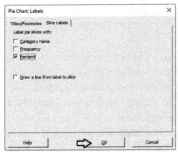

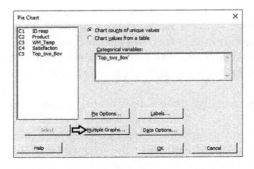

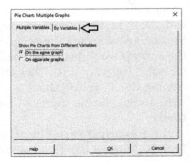

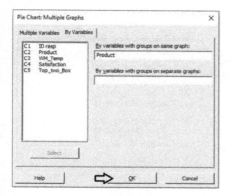

6.2.1.1.1 Interpret the Results of Step 1 The cross tabulation (Stat Tool 6.1) and the bar chart (Stat Tool 6.2) show the bivariate frequency distribution of the satisfaction score by product. The distribution of the satisfaction scores is J-shaped for both products (Stat Tool 1.5). The most common score, i.e. the mode (Stat Tool 1.4), is 5 with a percentage of 37.0% for product A and 46.4% for product B. 26.0% of respondents evaluating product A gave scores of 1 or 2 vs 11.9% for product B. Respondents' opinions vary less for product B (mode = 46.4%).

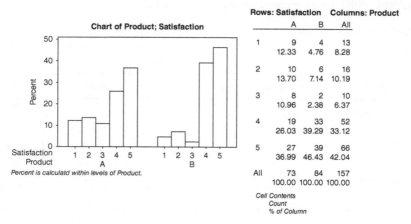

Tabulated Statistics: Satisfaction; Product

Rows: Satisfaction Columns: Product

	A	B	All
1	9	4	13
	12.33	4.76	8.28
2	10	6	16
	13.70	7.14	10.19
3	8	2	10
	10.96	2.38	6.37
4	19	33	52
	26.03	39.29	33.12
5	27	39	66
	36.99	46.43	42.04
All	73	84	157
	100.00	100.00	100.00

Cell Contents
Count
% of Column

Chart of Product; Satisfaction

Percent is calculatd within levels of Product.

For product B the percentage of respondents in the "top two box" is higher (85.7% vs 63.0%) and customers assigning overall satisfaction scores of less than 4 are less than half of those related to product A (14.3% vs 37.0%).

Tabulated Statistics: Top_two_Box; Product

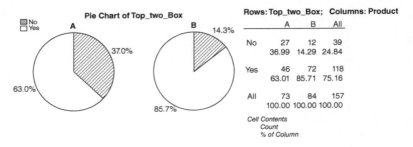

Pie Chart of Top_two_Box

Rows: Top_two_Box; Columns: Product

	A	B	All
No	27	12	39
	36.99	14.29	24.84
Yes	46	72	118
	63.01	85.71	75.16
All	73	84	157
	100.00	100.00	100.00

Cell Contents
  Count
  % of Column

Stat Tool 6.1 Cross Tabulations

Considering *two categorical variables*, a *cross tabulation* (also known as a *two-way table* or *contingency table*) shows the *bivariate frequency distribution* of the two variables. Note that you can also create a cross tabulation with *quantitative* variables, for example, with discrete variables assuming few values or continuous variables that have been transformed into categorical variables by aggregating values into few intervals.

A cross tabulation (Table 6.1) shows the *joint* frequency distribution (the central part of the table) and two *marginal* frequency distributions, one for each variable (the last column and the last row of the table).

Table 6.1 Cross-tabulation.

	Variable 2				
Variable 1	First category	Second category	...	Last category	Total
First category	...	...	...	...	...
Second category	...	...	...	...	...
...	...	...	...	...	...
Last category	...	...	...	...	...
Total	...	...	...	...	...

The *joint frequency distribution* shows the absolute or relative counts for each combination of categories of the two variables.

The two *marginal frequency distributions* show the absolute or relative counts for each variable, regardless of the other variable.

By moving into bivariate analysis, we start examining *relationships* between variables.

Stat Tool 6.1 (Continued)

➤ *Example 6.1.* In a consumer test, 88 evaluations were recorded related to the purchasing intentions of consumers with respect to a new formulation of an antiseptic spray. The cross-tabulation of purchase intent by gender was calculated (Table 6.2).

Table 6.2 Cross-tabulation of purchase intent by gender.

Purchase intent	Women		Men		Total	
	N	%	N	%	N	%
1) Definitely would not buy it	5	5.7	6	6.8	11	12.5
2) Probably would not buy it	6	6.8	5	5.7	11	12.5
3) Neither	8	9.1	5	5.7	13	14.8
4) Probably would buy it	14	15.9	19	21.6	33	37.5
5) Definitely would buy it	13	14.8	7	7.9	20	22.7
Total	46	52.3	42	47.7	88	100

- Joint distribution of purchase intent by gender (counts and percentages)
- Marginal distribution of purchase intent (counts and percentages)
- Marginal distribution of gender (counts and percentages)

In Table 6.2, the percentages are calculated, dividing the absolute counts by the overall total of consumers ($N = 88$). For different purposes, we can calculate the percentages with respect to each category of one of the two variables. For example, if we want to compare the distribution of purchase intent between men and women, removing the possible bias due to different totals, we can calculate the so-called *conditional* percentages, dividing the absolute counts by the total of men and by the total of women (Table 6.3). By doing this, we "condition" the study of purchase intent distribution to each category of gender.

Table 6.3 Counts and conditional percentages.

Purchase intent	Women		Men		Total	
	N	%	N	%	N	%
1) Definitely would not buy it	5	10.9	6	14.3	11	12.5
2) Probably would not buy it	6	13.0	5	11.9	11	12.5
3) Neither	8	17.4	5	11.9	13	14.8
4) Probably would buy it	14	30.4	19	45.2	33	37.5
5) Definitely would buy it	13	28.3	7	16.7	20	22.7
Total	46	100	42	100	88	100

- Distribution of purchase intent conditioned to women (counts and percentages)
- Distribution of purchase intent conditioned to men (counts and percentages)
- Unconditional or marginal distribution of purchase intent (counts and percentages)

Stat Tool 6.2 Bar Charts and Pie Charts

The *bivariate frequency distribution* of two variables can also be represented through graphs by using:

- *Bar charts*, where the bars' extents represent the frequencies or relative frequencies (usually percentages) of each category.
- *Pie charts*, displaying the proportion (percentages) of each category related to the total as slices of a pie.

Note that the difference between a bar chart and a histogram is that in a histogram the bars are adjacent and their bases represent intervals of continuous numeric values, while in a bar chart the bars are spaced and represent qualitative categories or discrete numbers.

By observing a bar chart or a pie chart, different *shapes* may be detected: uniform/nonuniform, unimodal/multimodal (Stat Tool 1.4). You can add comments about the presence of symmetry or skewness in the distribution (Stat Tools 1.1, 1.5) but only when the graph is related to an ordinal categorical or a quantitative discrete variable.

Remember that the category or discrete value with the highest frequency is the *mode* of the distribution.

The *percentage of the mode* can be used as a measure of variability for variables, especially when it is not possible to calculate other measures of variability such as range, interquartile range, or standard deviation (i.e. for nominal categorical variables):

- When categories have similar percentages and the distribution is fairly uniform, the statistical units are spread over categories. The percentage of the mode will be quite low, denoting *high variability* among units.

low percentage of the mode
high variability

- When the distribution is not uniform and the percentage of the mode is high, the statistical units are less spread over categories, denoting *low variability* among units.

high percentage of the mode
low variability

➤ *Example 6.1.* Consider the previous example (Stat Tool 6.1) and create the bar chart of purchase intent by gender (Figure 6.1). Both distributions are skewed to the left denoting higher frequencies of consumers with score 4 and 5. The mode of purchase intent is score 4 for both women and men, but the percentage of the mode is higher for men (45.2% vs. 30.4%) indicating less variability of the scores among this category of consumers.

Stat Tool 6.2 (Continued)

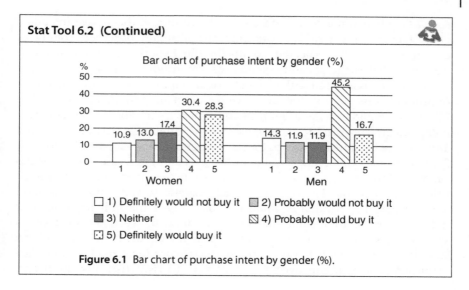

Figure 6.1 Bar chart of purchase intent by gender (%).

6.2.1.2 Step 2 – Apply the χ^2 Test to Evaluate the Presence of Association Between Product and Satisfaction Score

A preliminary analysis that we can apply to explore the relationship between the nominal variable "Product" and the ordinal variable "Top_two_Box," is the χ^2 test. We consider the variable "Top_two_Box" instead of "Satisfaction" to avoid expected counts that are too small as required by the test (Stat Tool 6.3).

To apply the χ^2 test, go to: **Stat > Tables > Chi-square Test for Association**.

In the drop-down menu at the top, choose **Raw data (categorical variables)**. In **Rows**, select "Top_two_Box" and in **Columns**, select "Product" and click **OK**.

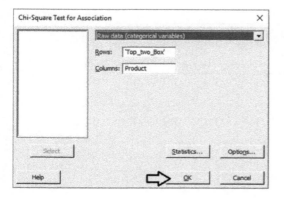

6.2.1.2.1 Interpret the Results of Step 2 The results of the χ^2 test include a cross tabulation where you can find the absolute observed counts of consumers and the expected counts in the hypothesis of independent variables, that is the absolute counts we would observe if the satisfaction score was not associated with product. For example, consider the consumers of product A in the "top-two box": we recorded an absolute count equal to 27, but we would observe around 18 consumers if the satisfaction score was independent from product, and so on. Are the differences between observed and expected counts in the hypothesis of independence statistically different from 0 or not? Look at the p-value of the χ^2 test. To determine whether the variables "Top_two_Box" and "Product" are independent, compare the p-value related to the *Pearson* chi-square statistic to the significance level $\alpha = 0.05$. As the p-value is equal to 0.001, reject the null hypothesis of independence and conclude that the association between the variables is statistically significant at a significance level of 0.05: the distribution of the satisfaction score tends to vary between the two products and the difference in the percentages of "top-two box" (85.7% vs 63.0%) is statistically significant.

Chi-square test for association: Top_two_Box; product

Rows: Top_two_Box; Columns: Product

	A	B	All
No	27	12	39
	18.13	20.87	
Yes	46	72	118
	54.87	63.13	
All	73	84	157

Cell Contents
  Count
  Expected count

Chi-square test

	Chi-square	DF	p-Value
Pearson	10.780	1	0.001
Likelihood Ratio	10.925	1	0.001

Stat Tool 6.3 Chi-square Test

A cross-tabulation displaying the joint frequency of data that are categorized by two categorical variables represents a *descriptive* tool:

- To compare percentages among categories;
- To investigate the relationship between the two variables.

After having explored these issues from a descriptive point of view, we can proceed to apply a hypothesis test (Stat Tools 1.3, 1.15, 1.16) to determine if any difference among categories is statistically significant or if there is a statistically significant association between variables.

Considering a *contingency table*, use the *chi-square test* to determine whether the two variables are related to each other or are independent (i.e. the variation of one variable does not influence the variation of the other).

For a χ^2 test:

Null Hypothesis H₀:
The variables are independent; no association exists between variables.

Alternative Hypothesis H₁:
The variables are not independent; an association exists between variables.

Setting the significance level α at usually 0.05 or 0.01:

- If the p-value is less than α, reject the null hypothesis and conclude that the association between variables is statistically significant.   p-value $< \alpha$ Reject the null hypothesis of independence of variables.
- If the p-value is greater than or equal to α, fail to reject the null hypothesis and conclude that the association between variables is NOT statistically significant.   p-value $\geq \alpha$ Fail to reject the null hypothesis of independence of variables.

The χ^2 test requires a *random sample* selected from a population (Stat Tool 1.2) with each observation independent of all other observations, i.e. what we measured from one unit does not influence the other measurements. Furthermore, the expected counts for each category must not be too small (generally, not less than 5). If need be, you can combine a category with low frequencies with adjacent categories to achieve the minimum expected count or use Fisher's exact test, which is accurate for all sample sizes.

➤ *Example 6.1.* Consider the example in Stat Tool 6.1. In the cross-tabulation (Table 6.4), for each category of purchase intent, and separately for women

Stat Tool 6.3 (Continued)

and men, you can find: the absolute counts, the conditional percentage of purchase intent by gender, and the expected counts in the hypothesis of independent variables, i.e. the absolute counts we would observe if purchase intent was not associated with gender. For example, consider the women who responded "Definitely would not buy it": we recorded an absolute count of 5 (10.87% of total women), but we would observe 5.75 women if the purchase intent was independent from gender, and so on. Are the differences between observed and expected counts in the hypothesis of independence statistically different from 0 or not? Look at the p-value of the chi-square test. The p-value is equal to 0.516, greater than the significance level $\alpha = 0.05$. We fail to reject the null hypothesis. The association between purchase intent and gender is not statistically significant.

Table 6.4 Cross-tabulation of purchase intent by gender and chi-square test results.

	Women	Men	All
1) Definitely would not buy it	5	6	11
	10.87	14.29	12.50
	5.750	5.250	
2) Probably would not buy it	6	5	11
	13.04	11.90	12.50
	5.750	5.250	
3) Neither	8	5	13
	17.39	11.90	14.77
	6.795	6.205	
4) Probably would buy it	14	19	33
	30.43	45.24	37.50
	17.250	15.750	
5) Definitely would buy it	13	7	20
	28.26	16.67	22.73
	10.455	9.545	
All	46	42	88
	100.00	100.00	100.00

Cell contents

Count	Chi-square	DF	p-Value
% of column	Pearson	3.257 4	0.516
Expected count			

Tabulated Statistics: Purchase intent; Gender
Rows: Purchase intent; Columns: Gender

6.2.1.3 Step 3 – Perform an Exploratory Descriptive Analysis of Variable "Satisfaction" by Temperature and Calculate the Correlation Coefficient

For the discrete variable "Satisfaction," use the bar chart or pie chart, stratifying by washing machine temperature, to show separate satisfaction score distributions for each temperature. Add the frequency tables to complete the descriptive analysis.

To display the bar chart of "Satisfaction" by temperature, go to: **Graph > Bar Chart**.

Select the graphical option **Cluster** and click **OK**. In the next dialog box, in **Categorical variables**, choose the stratification variable "WM_temp" and the response "Satisfaction." Click **Chart Options**. Under **Percent and Accumulate**, check the option **Show Y as Percent**, and under **Take Percent and/or Accumulate**, check **Within categories at level 1**. Click **OK** in each dialog box.

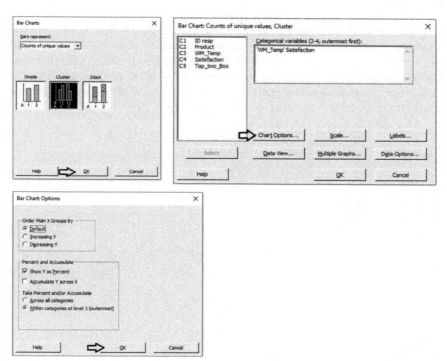

To display the frequency table of "Satisfaction" by temperature, go to: **Stat > Tables > Cross Tabulation and Chi-square**

In **Rows**, select "Satisfaction" and in **Columns**, select "WM_temp." Under **Display**, check **Counts** and **Column percents** and click **OK**.

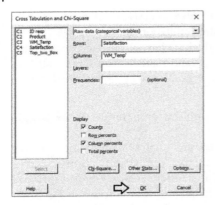

To display the bar chart of "Top_two_Box" by temperature, go to: **Graph > Bar Chart**

Select the graphical option **Stack** and click **OK**. In the next dialog box, under **Categorical variables**, choose the stratification variable "WM_temp" and the response "Top_two_Box." Click **Chart Options**. Under **Percent and Accumulate**, check the option **Show Y as Percent**, and under **Take Percent and/or Accumulate,** check **Within categories at level 1**. Click **OK** in each dialog box. Instead of a bar chart, you can use a pie chart as we did for the graph of the variable "Top_two_Box" by product.

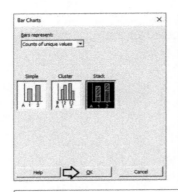

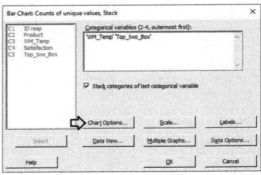

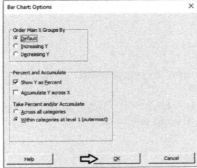

To calculate the coefficient of correlation between the satisfaction score and the temperature, go to: **Stat > Basic Statistics > Correlation**

Select the variables "WM_temp" and "Satisfaction." In **Method**, choose **Spearman rho**. Check the option **Display p-values**, and click **OK**.

6.2.1.3.1 *Interpret the Results of Step 3*

The distribution of the satisfaction scores is J-shaped for all three temperatures. The mode is score 5 with percentages of 34.0% for 30 °C, 44.9% for 40 °C, and 50.0% for 60 °C. No consumer made negative judgments (scores equal to 1 or 2) using the product at a temperature of 60 °C. Note that only 18 consumers tested the product at 60 °C. Respondents' opinions vary less for the higher temperature (percentage of the mode = 50.0%), while more spread is present for the lower temperature (percentage of the mode = 34%). For 30 °C the percentage of respondents in the "top two box" is 58%, while for 40 and 60 °C the "top two box" percentage for both is just over 83%.

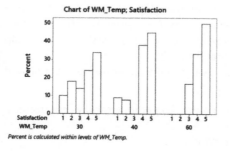

Tabulated Statistics: Satisfaction; WM_Temp

Rows: Satisfaction   Columns: WM_Temp

	30	40	60	All
1	5	8	0	13
	10.00	8.99	0.00	8.28
2	9	7	0	16
	18.00	7.87	0.00	10.19
3	7	0	3	10
	14.00	0.00	16.67	6.37
4	12	34	6	52
	24.00	38.20	33.33	33.12
5	17	40	9	66
	34.00	44.94	50.00	42.04
All	50	89	18	157
	100.00	100.00	100.00	100.00

Cell Contents
   Count
   % of Column

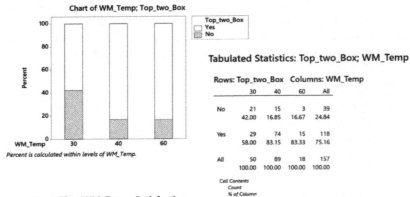

Chart of WM_Temp; Top_two_Box

Percent is calculated within levels of WM_Temp.

Tabulated Statistics: Top_two_Box; WM_Temp

Rows: Top_two_Box Columns: WM_Temp

	30	40	60	All
No	21	15	3	39
	42.00	16.85	16.67	24.84
Yes	29	74	15	118
	58.00	83.15	83.33	75.16
All	50	89	18	157
	100.00	100.00	100.00	100.00

Cell Contents
    Count
    % of Column

Spearman Rho: WM_Temp; Satisfaction

Correlations
Spearman rho 0.188
P-value   0.018

The correlation coefficient (Spearman rank correlation) between temperature and satisfaction scores is 0.19, which indicates that the two variables have a weak positive relationship (Stat Tool 6.4). Setting the significance level α at 0.05, the p-value (0.018) is less than 0.05. The relationship between the two variables is, however, statistically significant.

Stat Tool 6.4 Spearman Rank Correlation

A correlation coefficient helps to measure the relationship between two variables describing how they tend to change together.

In the previous chapter (Stat Tool 5.3) we learned how to calculate and interpret the Pearson correlation. This correlation coefficient quantifies the *linear* association between two *quantitative* variables. Also remember that a scatterplot is the first descriptive tool to use in evaluating a possible relationship between two quantitative variables (Stat Tool 5.2).

Instead of the Pearson correlation, we can use the Spearman rank correlation when:

- The relationship between the two variables is *monotonic* (always not decreasing or not increasing) but not necessarily linear.
- One or both variables are *ordinal categorical* variables.

Spearman's correlation coefficient is a single value ranging from minus one to plus one.

Stat Tool 6.4 (Continued)

The *sign* of the coefficient indicates the *direction* of the relationship:

- If one variable tends to increase as the other decreases, the coefficient is *negative*.
- If one variable tends to increase as the other increases, the coefficient is *positive*.
- If there is no monotonic relationship between variables, the coefficient equals 0.

The *magnitude* of the coefficient indicates the *strength* of the association:

- For a *stronger* relationship, the correlation is closer to plus one or minus one.
- A correlation near zero indicates a weak relationship (Figure 6.2).

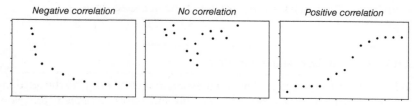

Figure 6.2 Scatterplots.

After calculating the correlation coefficient using **sample data**, we can apply a **hypothesis test** (Stat Tools 1.15–1.16) to determine whether a monotonic relationship between the variables exists. Let's denote the Spearman correlation by r for sample data and by the Greek letter ρ (rho) for population. For this test, the null and the alternative hypotheses are the following:

Null Hypothesis:	**Alternative Hypothesis**:
No monotonic correlation exists between the variables.	A monotonic correlation exists between the variables.
$H_0: \rho = 0$	$H_1: \rho \neq 0$

Setting the significance level α at usually 0.05 or 0.01:

- If the p-value is less than α, reject the null hypothesis and conclude that the correlation between the two variables is statistically significant.

p-value $< \alpha$ Reject the null hypothesis that the population correlation coefficient equals 0.

Stat Tool 6.4 (Continued)

• If the p-value is greater than or equal to α, fail to reject the null hypothesis and conclude that the correlation between the two variables is NOT statistically significant.	p-value $\geq \alpha$	Fail to reject the null hypothesis that the population correlation coefficient equals 0.

➢ *Example 6.1.* Consider the example in Stat Tool 6.1. Investigators also recorded the age of respondents categorized in four classes: (i) 20–29; (ii) 30–39; (iii) 40–49; (iv) ≥ 50 years. Purchase intent and age are two ordinal categorical variables. The Spearman rank correlation between them is equal to −0.73 with a p-value of 0.0001. The correlation between the two variables is statistically significant. The purchase intent tends to be lower as age increases.

Stat Tool 6.5 Logistic Regression Models

So far, we have learned to use linear regression models to examine the relationship between a quantitative response variable and one or more predictors. Now we introduce *logistic regression models* to analyze how one or more explanatory variables are related to a *categorical* (Stat Tool 1.1) response (Figure 6.3).

Logistic regression models include the following:

• *Binary* logistic models: the response is a *binary* variable assuming only two categories or outcomes that are usually denoted as *success* or *failure*, such as presence and absence of defects.
• *Ordinal* logistic models: the response is an *ordinal* categorical variable with more than two categories that have an order, such as low satisfaction, medium satisfaction, and high satisfaction.

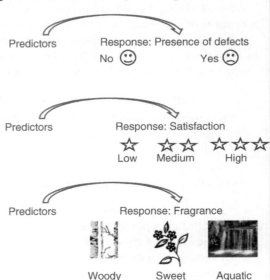

Figure 6.3 Different categorical response variables.

Stat Tool 6.5 (Continued)	

• *Nominal* logistic models: the response is a *nominal* categorical variable with more than two categories that do not have an order, such as woody fragrance, sweet fragrance, aquatic fragrance.

In logistic regression, predictors can be quantitative variables (continuous or discrete) or categorical.

Be aware of the correlation among the predictors, also known as *multicollinearity*. If multicollinearity exists and is severe, you might not be able to estimate the contribution of each predictor to the response. To check for multicollinearity, examine the correlation (Stat Tools 5.3, 6.4) between the predictor variables before creating the model.

The steps to follow in a logistic regression analysis are similar to those we presented earlier for linear regression models:

1) Fit the logistic model by estimating one coefficient (slope) for each explanatory variable.
2) Check the statistical significance of each coefficient (slope) by hypothesis testing.
3) Check the goodness of fit of the model (Arboretti Giancristofaro, R. and Salmaso, L., 2003).

6.2.1.4 Step 4 – Fit a Binary Logistic Regression Model of Satisfaction Score vs. Temperature and Product

Since the research team is particularly interested in the percentage of respondents belonging to the "top two box" (that is customers assigning scores 4 and 5 to overall satisfaction), we complete our analysis by building a binary logistic model in which the variable "Top_two_Box" is the response and washing machine temperature and product are the explanatory variables.

 To fit a binary logistic regression model, go to: **Stat > Regression > Binary logistic Regression > Fit Binary Regression Model**

In the drop-down menu at the top, choose **Response in binary response/frequency format**, then select the binary response variable "Top_two_Box" in **Response**. In **Response event**, select which event the analysis will describe, for example the value "Yes," taking into account that changing the response event does not affect the overall significance of the results. Under **Categorical predictors**, select "WM_temp" and "Product" and click on **Model**. By default, the model contains only the main effects for the predictor variables specified in the main dialog box. However, you can add, for example, a two-way interaction term (i.e. an interaction between two predictors) to your model; use the mouse to highlight a couple of explanatory variables and next to **Interactions through order: 2,** select **Add**. In the present example we don't include the product/temperature interaction term as investigators consider their interaction to be irrelevant, based also on the results of previous studies. Click **OK** in each dialog box.

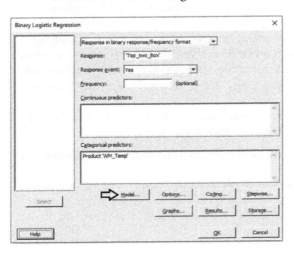

6.2.1.4.1 *Interpret the Results of Step 4*   In the Deviance table we examine the p-values to determine whether any predictor or interaction is statistically significant. Remember that the p-value is a probability that measures the evidence against the null hypothesis. Lower probabilities provide stronger evidence against the null hypothesis.

For a main effect, the null hypothesis is that there is no association between a predictor and the response. For an interaction term, H_0 states that the relationship between the response and one of the predictors involved in the interaction does not depend on the other predictors in the term.

Usually we consider a significance level alpha equal to 0.05, but in an exploratory phase of the analysis we may also consider a significance level of 0.10.

Response information

Variable	Value	Count
Top_two_Box	Yes	118 (Event)
	No	39
	Total	157

Deviance table

Source	DF	Adj dev	Adj mean	Chi-square	P-value
Regression	3	20.524	6.841	20.52	0.000
Product	1	9.486	9.486	9.49	0.002
WM_Temp	2	9.599	4.800	9.60	0.008
Error	153	155.497	1.016		
Total	156	176.022			

When the p-value is greater than or equal to alpha, we fail to reject the null hypothesis. When it is less than alpha, we reject the null hypothesis and claim statistical significance.

So, setting the significance level alpha at 0.05, which terms in the model are significant in our example? The answer is: Both main effects of temperature and product are statistically significant.

Bear in mind that if you have several predictors in your model and some of the effects are not statistically significant, you may want to reduce your model by using an automatic selection procedure such as the *stepwise strategy* (see Chapter 5 for an example in linear regression models).

The interpretation of a statistically significant term in the binary logistic model depends on the type of term:

- For a continuous predictor, you can conclude that the regression coefficient for the predictor is different from zero, denoting a significant association with the response.
- For a categorical predictor, you can conclude that not all of the categories of the predictor have the same probability of the response event.
- For an interaction term, you can conclude that the relationship between a predictor and the probability of the response event depends on the other predictors in the term.

In these results, the product is statistically significant at the significance level of 0.05 (p-value = 0.002). You can conclude that the two products are associated with different probabilities of the event occurring. Remember that the event is that the opinion of a consumer is in the "top two box." Also in relation to washing machine temperature, as the term is statistically significant (p-value = 0.008), you can conclude that different temperatures are associated with different probabilities of the event "being in the top two box."

To better understand the effects of the predictors on the response, let us use the information given by the *odds ratios* (Stat Tool 6.6). In our example, the event is "being in the top two box"; then the odds are the probability of "being in the top two box" divided over the probability of "not being in the top two box."

With regard to the product, the odds ratio equals to 3.39 with a 95% CI, not including the value "1" means that product B shows odds of the event more than three times greater than product A (the reference level).

Odds ratios for categorical predictors

Level A	Level B	Odds ratio	95% CI
Product			
B	A	3.3865	(1.5196; 7.5470)
WM_Temp			
40	30	3.4961	(1.5428; 7.9228)
60	30	3.0460	(0.7486; 12.3938)
60	40	0.8713	(0.2152; 3.5282)

With regard to washing machine temperature, we can see that only the odds ratio comparing the 40 °C level to the 30 °C level (reference level) is statistically significant (its 95% CI does not include the value "1"), showing odds of the event for 40 °C more than three times greater than for 30 °C.

Stat Tool 6.6 Odds Ratios

To better understand the effects of the predictors on the response, you can use the information given by a measure of association called "odds ratio."

The *odds* of an *event* are the probability that the event occurs divided over the probability that the event does not occur.

Odds of an event = Probability that the event occurs / Probability that the event does not occur

➤ *Example 6.1.* Consider the example in Stat Tool 6.1 and suppose that the researchers are particularly interested in consumers answering that they definitely or probably would buy the product. A purchase intent equal to "definitely / probably would buy it" is the *event* of interest.

Let's create a new variable Y that assumes the value "1" when the purchase intent is equal to 4 or 5 (definitely/probably would buy it), and "0" when the purchase intent is equal to 1, 2, or 3.

For our *event* of interest the odds are the probability of "1" divided over the probability of "0":

Odds of "definitely / probably would buy it" = (Probability that $Y = 1$) / (Probability that $Y = 0$)

Consider now a *categorical* predictor and imagine calculating the *odds* of an event for two different levels of the predictor, say Level A and Level B. If we compute the ratio of the two odds, we obtain the *odds ratio* for that predictor.

Stat Tool 6.6 (Continued)

The level with odds in the denominator of the odds ratio is called the *reference level* for the predictor. The interpretation of an odds ratio is the following:

- Odds ratio > 1: the event is less likely at the reference level than at the other level.
- Odds ratio < 1: the event is more likely at the reference level.
- Odds ratio = 1: the event is equally likely for both levels.

Odds ratios are fairly easy to interpret when they are greater than one. For example, an odds ratio equal to 1.44 means that the level of the predictor in the numerator of the odds ratio has a 44% increase in the odds of the event than the reference level. Odds ratios are less easily grasped when the value is less than one. To interpret an odds ratio of less than 1, it is worth calculating the inverse of its value. The obtained value, greater than one, corresponds to the odds ratio above which has the reference level in the numerator and the other level in the denominator, thus representing the increase in the odds of the event that characterized the previous reference level with respect to the other level.

For example, suppose that the odds ratio relative to the event "preference for a liquid detergent rather than a powdered detergent" for the predictor "gender of the consumer" with the category "female" as the reference level, is equal to 0.83. The inverse 1/0.83 = 1.20 represents the odds ratio of the event where the reference level is now the category "male," meaning that women show a 20% increase in the odds of the event versus men.

The estimate of an odds ratio is generally shown with its 95% confidence interval. The confidence interval (CI) is a range of values that is likely to contain the true values of the odds ratio. For example, with a 95% confidence level, you can be 95% confident that the confidence interval contains the value of the odds ratio for the population. Be careful with the interpretation of an odds ratio when its CI includes the value "1." In this case, you don't have enough evidence to exclude that the event is equally likely for the two levels of the predictor.

If the predictor is a *quantitative* variable, odds ratios that are greater than 1 indicate that the event is more likely to occur as the predictor increases by one unit, while odds ratios that are less than 1 indicate that the event is less likely to occur as the predictor increases.

➤ *Example 6.1.* For the example in Stat Tool 6.1, consider the gender and age (as a continuous variable) of the respondents as two predictors. The researchers need to establish whether a favorable opinion on the purchase intent is more likely among women or men and as age increases or decreases. Through

Stat Tool 6.6 (Continued)

a binary logistic regression analysis (Stat Tool 6.5) with the variable "Y" as the response and its value "1" (definitely/probably would buy it) as the event, they calculate the odds ratio of gender taking "female" as the reference level (Level B in the results table) and the odds ratio of age (Table 6.5).

Table 6.5 Odds ratios.

Odds ratios for continuous predictors

	Odds ratio	95% CI
Age	0.9069	(0.8645; 0.9514)

Odds ratios for categorical predictors

Level A	Level B	Odds ratio	95% CI
Gender			
M	F	0.7656	(0.2862; 2.0479)

Looking at the odds ratio (0.77) for gender, it seems that the odds of a male consumer showing a favorable purchase intent is less than the odds for a female consumer. Computing the inverse $1/0.77 = 1.30$ shows that women have a 30% increase in the odds of the event than men. But to definitively evaluate whether or not this association is statistically significant, look at the 95% confidence interval.

Since the 95% CI (0.29–2.05) includes the value "1," the researchers can't exclude the possibility of the odds of expressing favorable purchase intent being equally likely for males and females.

Consider now the odds ratio for the continuous variable Age: 0.91. It seems that the odds of a favorable purchase intent decrease as age increases and since the 95% CI (0.86–0.95) does not include the value "1," this result is statistically significant. By computing the inverse $1/0.91 = 1.10$, the researchers can conclude that as age decreases by one year, the odds of the event grow by 10%.

The results obtained by estimating odds ratios and related 95% CI through binary logistic analysis, confirm the previous outcomes derived from the application of the chi-square test (Stat Tool 6.3) for gender, and from the calculation of the Spearman rank correlation (Stat Tool 6.4) for age (split into four classes).

In general, the approach based on a regression model that looks for an association between a response and *more than one categorical or continuous predictor*, allows us to explore the contribution of each predictor taking into account or "adjusting" for the effects of the other predictors included in the model. This is why in a binary logistic regression model with more than one predictor, the odds ratios are said to be "adjusted odds ratios." Results obtained from chi-square tests or correlation coefficients (Pearson correlation or Spearman rank

> **Stat Tool 6.6 (Continued)**
>
> correlation) are related to the study of the association between a response and a *single predictor*. If the researcher suspects that other predictors could be related to the response, it is better to continue the analysis by looking for a regression model which includes all the predictors to take into account possible confounding effects among them. When a predictor alters or distorts the relationship between another predictor and the response, we call this a confounding effect. The estimated model could confirm the previous results or show different trends in the association between response and predictors when actually the latter are the confounders.

In addition to the results shown in the deviance table, Minitab displays some other useful information in the Model Summary table to evaluate how well the model fits the data.

The quantity **deviance R-squared** (R-sq, R^2) has a similar interpretation as the coefficient of determination R^2 in linear regression (Stat Tool 5.6). The value of deviance R-sq varies from 0% to 100%, with larger values being more desirable.

Model summary

Deviance R-Sq	Deviance R-Sq(adj)	AIC
11.66%	9.96%	163.50

In our example, the model explains only 11.7% of the deviance in the response variable. For these data, the deviance R^2 value indicates that the model could be improved, for example, by adding other predictors to provide a better fit to the data.

Note that the deviance R^2 always increases when you add additional predictors to a model. Therefore, use deviance R^2 to compare models that include the same number of explanatory variables. The deviance adjusted R^2 (R-sq(adj)) is a variation of the ordinary deviance R^2 that is adjusted for the number of terms in the model. Use deviance adjusted R^2 when you want to compare several models with different numbers of terms.

Additionally, the value of AIC **(Akaike Information Criteria)** may be used to compare different models. Generally, the smaller it is, the better the fit of the model to the data.

The model's goodness-of-fit can also be checked through several hypothesis tests. For the interpretation of these tests, pay attention to the format of the data in the worksheet for the response and predictors. When we built the logistic model, we selected the option **Response in binary response/frequency format**, since our data has a column reporting the value "Yes" or "No" for the "Top_two_Box" response for each consumer. In this case, we can use both the deviance test and the Hosmer-Lemeshow test, while the Pearson test is not adequate. For all these tests, the null hypothesis states that the model predicts well the probabilities of the response event.

In our results, both tests show a p-value greater than the usual significance level of 0.05. So we fail to reject the null hypothesis and conclude that there is no evidence that the model does not predict well the probabilities of the event.

Goodness-of-fit tests

Test	DF	Chi-square	P-value
Deviance	153	155.50	0.429
Pearson	153	158.58	0.362
Hosmer-Lemeshow	2	0.49	0.782

6.3   Case Study: DOE – Top Score Project

In chapters 2 and 3, factorial designs were created and continuous response variables were analyzed through analysis of variance and response surface methodology. Sometimes it may be of interest to add more information on the relationship between factors and responses by transforming the latter into binary response variables and applying binary logistic regression models as in the previous case study 6.2.

During a previous screening and development phase, investigators selected two additives A and B significantly affecting the technical performance of a laundry detergent.

Now, considering two levels for each additive (absence/presence), the researchers want to run a consumer test on the four different formulations of the laundry detergent given by the combinations of the two additives. A random sample of 200 consumers will test the four different formulations for a month and at the end of the period the overall opinion on the product will be recorded with a score from 1: dislike extremely, to 10: like extremely.

The main purpose is to evaluate how the different levels of additives affect the probability of a liking score being greater than or equal to 8, denoted as "top score."

6.3.1   Plan of the Factorial Design

For the present study:

- The *two additives* represent the *key factors*.
- Each factor has two *levels* (absence and presence).
- The *liking scores* represent *the response variable*.

The researchers proceed by constructing a 2^2 factorial design in which all four combinations of factor levels will be tested. "-1" and "$+1$" represent the two levels of each additive, and the basic design includes the following combinations:

A	B	Treatment/Combination (Product formulation)
-1	-1	1. A absent, B absent
$+1$	-1	2. A present, B absent
-1	$+1$	3. A absent, B present
$+1$	$+1$	4. A present, B present

Each formulation is then randomly assigned to 50 people with a total of 200 different consumers. At the end of the one-month trial period, during which detailed usage instructions are given to the consumers, data about gender, age, type of garments washed (white or colored), and other usage details are recorded along with the relevant liking score.

6.3.2 Plan of the Statistical Analyses

In order to study in depth how additives A and B are associated with liking scores greater than or equal to 8, we will fit a binary logistic regression model by transforming the liking score (from 1 to 10) into a binary response variable "Top score" assuming value "No" when the liking score is less than 8 and value "Yes" when the liking score is greater than or equal to 8.

Let's proceed in the following way:

Step 1 – Perform a descriptive analysis (Stat Tool 1.3) of the binary response variable "Top_score" stratifying by formulations.
Step 2 – Fit the binary logistic regression model.
Step 3 – If need be, reduce the model to include the significant terms and to estimate the odds ratios for additives A and B.

The variables setting of the dataset extracted from the study is the following:

- Variables Additive_A, Additive_B, are the categorical factors, each assuming 2 levels.
- Variable Top_score is the binary response variable, assuming the value "No" when the liking score is less than 8, and the value "Yes" otherwise.

Column	Variable	Type of data	Label
C1	ID_consumer		Consumer's code
C2	Additive_A	Categorical data	Additive A with levels: −1 (absent), +1 (present)
C3	Additive_B	Categorical data	Additive B with levels: −1 (absent), +1 (present)
C4	Formulation	Categorical data	Product formulation (combinations of additives)
C5	Gender	Numeric data	Consumer's gender: 1 (male), 2 (female)
C6	Age	Categorical data	Consumer's age
C7	Type_of_garment	Categorical data	Type of garments (white/colored)
C8	Top_score	Categorical data	Liking score less than 8 (NO), Otherwise (YES)

DOE_Top_score_Project.xlsx

Note that the dataset includes three variables: Gender, Age, and Type of Garments (white/colored). This takes into account possible effects on the response due to general characteristics of the consumers and usage conditions.

6.3.2.1   Step 1 – Perform a Descriptive Analysis of the Binary Response Variable Stratifying by Formulations

For the categorical variable "Top_score," use the bar chart or the pie chart stratifying by formulations to show the distribution of the binary response separately for each formulation. Add the frequency tables to complete the descriptive analysis.

To display the bar chart of "Top_score" by formulations, go to: **Graph > Bar Chart**.

Select the graphical option **Stack** and click **OK**. In the next dialog box, under **Categorical variables**, choose the stratification variable "Formulation" and the response "Top_score." Click **Chart Options**. Under **Percent and Accumulate,** check the option **Show Y as Percent**, and under **Take Percent and/or Accumulate,** check **Within categories at level 1**. Click **OK** in each dialog box.

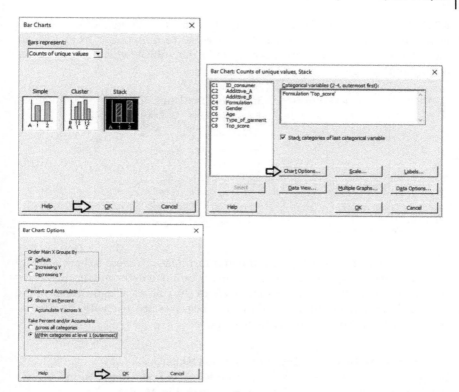

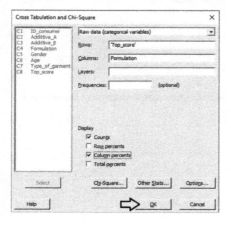

To display the frequency table of "Top_score" by formulations, go to: **Stat >
Tables > Cross Tabulation and Chi-square**.

In **Rows**, select "Top_score" and in **Columns**, select "Formulation." Under
Display, check **Counts** and **Column percents** and click **OK**.

6.3.2.1.1 Interpret the Results of Step 1 The percentage of consumers expressing a liking score greater than or equal to 8 increases from 12% for formulation 1 to 36% for formulation 3. For formulation 4 (where both additives are present) the percentage is 32%.

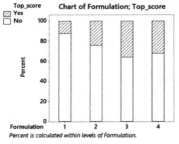

Chart of Formulation; Top_score

Percent is calculated within levels of Formulation.

Tabulated Statistics: Top_score; Formulation

Rows: Top_score   Columns: Formulation

	1	2	3	4	All	Cell Contents
						Count
No	44	38	32	34	148	% of Column
	88	76	64	68	74	
Yes	6	12	18	16	52	
	12	24	36	32	26	
All	50	50	50	50	200	
	100	100	100	100	100	

From a *descriptive point of view*, the four formulations seem to have a different effect on the response, and the presence of additive B (formulation 3) seems to be associated with the best result. To check the statistical significance of the relationship between additives and the response, let's proceed with the inferential analysis based on the binary logistic model.

6.3.2.2   Step 2 – Fit a Binary Logistic Regression Model

To fit a binary logistic regression model, go to: **Stat > Regression > Binary logistic Regression > Fit Binary Regression Model**.

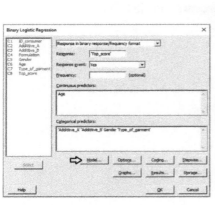

In the drop-down menu at the top, choose **Response in binary response/ frequency format**, then select the binary response variable "Top_score" in **Response**. In **Response event**, select which event the analysis will describe, e.g. the value "Yes," taking into account that changing the response event does not affect the overall significance of the results. Under **Continuous predictors**, select "Age," and under **Categorical predictors**, select "Additive_A," "Additive_B," "Gender" and "Type_of_garment." Then click on **Model**. By default the model contains only the main effects for the predictors specified in the main dialog box; however, you can for example add the two-way interaction between additives A and B by selecting the two variables "Additive_A" and "Additive_B" using the mouse and next to **Interactions through order: 2** choosing **Add**. Click **OK** in each dialog box.

6.3.2.2.1 *Interpret the Results of Step 2* In the deviance table we examine the p-values to determine whether any predictor or interaction is statistically significant. Setting the significance level alpha to 0.05, only the main effects of additive B is statistically significant. You may now want to reduce the model to include only significant terms.

Deviance table

Source	DF	Adj dev	Adj mean	Chi-square	p-Value
Regression	6	13.369	2.2281	13.37	0.038
Age	1	1.670	1.6702	1.67	0.196
Additive_A	1	2.816	2.8162	2.82	0.093
Additive B	1	9.441	9.4413	9.44	0.002
Gender	1	0.711	0.7112	0.71	0.399
Type_of_garment	1	2.716	2.7158	2.72	0.099
Additive_A*Additive_B	1	2.924	2.9238	2.92	0.087
Error	193	215.854	1.1184		
Total	199	229.223			

Remember that when you use statistical significance to decide which terms to keep in a model, it is usually advisable not to remove entire groups of non-significant terms at the same time. The statistical significance of individual terms can change because of other terms in the model. To reduce your model, you can use an automatic selection procedure, the *stepwise strategy*, to identify a useful subset of terms, choosing one of the three commonly used alternatives (standard stepwise, forward selection and backward elimination).

6.3.2.3 Step 3 – If Need be Reduce the Model to Include the Significant Terms and Estimate the Odds Ratios

To reduce the model, go to:

Stat > Regression > Binary logistic Regression > Fit Binary Regression Model

Retain the selections made in the previous step and in the main dialog box, choose **Stepwise**. In **Method** select **Backward elimination** and in **Alpha value to remove** keep the default value 0.1. Click **OK** in each dialog box.

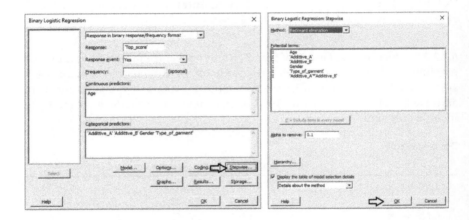

6.3.2.3.1 Interpret the Results of Step 3 In the deviance table we examine the p-values to determine whether any predictor or interaction is statistically significant. Setting the significance level alpha to 0.05, only the main effects of additive B are statistically significant. You can conclude that the probability of a liking score greater than or equal to 8 is statistically different for the two levels of additive B (absence/presence).

To better understand the effects of the predictor on the response, let's use the information given by the *odds ratios* (Stat Tool 6.6). In our example, the event is "a liking score greater than or equal to 8." The odds are the probability of "liking score ≥ 8" divided over the probability of "liking score < 8."

Considering additive B, the odds ratio equal to 2.35 with a 95% CI not including the value "1," means that the level coded "+1" of additive B (corresponding to the presence of the additive) shows odds of the event more than two times greater than the level coded "−1" (reference level: absence of the additive). The presence of additive B tends to be associated with a more favorable opinion of the product.

Deviance table

Source	DF	Adj dev	Adj mean	Chi-square	p-Value
Regression	1	6.737	6.737	6.74	0.009
Additive_B	1	6.737	6.737	6.74	0.009
Error	198	222.486	1.124		
Total	199	229.223			

Odds ratios for categorical predictors

Level A	Level B	Odds ratio	95% CI
Additive_B			
1	−1	2.3468	(1.2168; 4.5263)

Odds ratio for level A relative to level B

Goodness-of-fit tests

Test	DF	Chi-square	p-Value
Deviance	198	222.49	0.112
Pearson	198	200.00	0.447
Hosmer-Lemeshow	8	4.32	0.827

The model's goodness-of-fit can be checked by looking at the deviance test and the Hosmer-Lemeshow test. For all these tests, the null hypothesis states that the model predicts well the probabilities of the response event. In our results, both tests show a p-value greater than the usual significance level of 0.05. So we fail to reject the null hypothesis and conclude that there is no evidence that the model does not predict well the probabilities of the event.

6.4 Final Remarks

The binary logistic model finds wide application in chemical and engineering experiments with binary response data of the type success–failure, in biomedical research with dose-response data, and in consumer tests where the response is defined as whether or not the consumer gives satisfactory opinions.

In customer satisfaction evaluation, the combination of DOE techniques and logistic models can augment the possibility of researchers being able to thoroughly evaluate the impact of each factor level on the response. It also takes into account the possible effect of other predictors potentially related to the response and representing general characteristics of the consumers and environmental conditions, as shown in case study 6.3. In this context, the estimation of odds ratios and probabilities of the response event allows the researcher to quantify the relationship between factor levels and the response, even when stratifying for consumer characteristics or usage conditions. Let us now consider the following example.

For a customer satisfaction study on a new home air freshener product, a 2^2 design considering two factors (fragrance, type of diffuser), each with two levels, was created and tested by a random sample of consumers who assigned a satisfaction score of 1 (fully dissatisfied) to 10 (fully satisfied). Some general characteristics of the consumers were also recorded (age, gender, presence of children, etc.).

A binary logistic regression model was fitted to evaluate the relationship between factors and the probability of a satisfaction score greater than 8. In the model, variables representing consumer characteristics were also included.

At the end of the analysis the two factors (fragrance, type of diffuser) and the presence of children (with levels: yes/no) proved to be statistically significant. Along with the estimation of odds ratios, since the model included only categorical variables, the research team used the logistic model to estimate the probability (expressed in %) of the response event "a satisfaction score ≥ 8" for the eight consumer profiles. The results were displayed in a sort of "satisfaction card."

	No children			Presence of children	
	Type of diffuser 1	Type of diffuser 2		Type of diffuser 1	Type of diffuser 2
Fragrance A	14.3%	16.6%	Fragrance A	20.2%	23.2%
Fragrance B	27.8%	31.6%	Fragrance B	36.9%	41.2%

Based on these results, the researchers identified the most appealing factor levels for consumers and found information about different possible configurations of the product for different market segments.

Note that in Minitab you can visualize the fitted values of the response event probabilities by using the option "Storage" and then "Fits (event probabilities)" in the main dialog box of the binary logistic models.

References

Anderson, M.J. and Whitcomb, P.J. (2014). Practical aspects for designing statistically optimal experiments. *Journal of Statistical Science and Application* 2 (3): 85–92.

Anderson, M.J. and Whitcomb, P.J. (2015). *DOE Simplified: Practical Tools for Effective Experimentation*. Boca Raton, FL: CRC Press.

Arboretti Giancristofaro, R. and Salmaso, L. (2003). Model performance analysis and model validation in logistic regression. *Statistica* LXIII (2): 375–396.

Box, G.E.P. and Woodall, W.H. (2012). Innovation, quality engineering, and statistics. *Quality Engineering* 24 (1): 20–29.

Box, G.E.P., Hunter, J.S., and Hunter, W.G. (2005). *Statistics for Experimenters: Design, Innovation, and Discovery*. Hoboken, NJ: Wiley.

Cintas, P.G., Almagro, L.M., and Llabrés, X.T. (2012). *Industrial Statistics with Minitab*. Hoboken, NJ: Wiley.

Draper, N.R. and Pukelsheim, F. (1996). An overview of design of experiments. *Statistical Papers* 37 (1): 1–32.

Greenfield, T. and Metcalfe, A. (2007). *Design and Analyse Your Experiment with Minitab*. Hoboken, NJ: Wiley.

Hoerl, R. and Snee, R. (2010). Statistical thinking and methods in quality improvement: a look to the future. *Quality Engineering* 22 (3): 119–129.

Jensen, W., Anderson-Cook, C.M., Costello, J.A., and Doganaksoy, N. (2012). Statistics to facilitate innovation: a panel discussion. *Quality Engineering* 24 (1): 2–19.

Johnson, R.T., Montgomery, D.C., and Jones, B.A. (2011). An expository paper on optimal design. *Quality Engineering* 23 (3): 287–301.

Kounias, S. and Salmaso, L. (1998). Orthogonal plans of resolution IV and V. *Statistical Methods and Applications* 7 (1): 57–75.

Mathews, P.G. (2005). *Design of Experiments with Minitab*. Milwaukee, WI: ASQ Quality Press.

Montgomery, D.C. (2008). Applications of design of experiments in engineering. *Quality and Reliability Engineering International* 24 (5): 501–502.

Montgomery, D.C. (2017). *Design and Analysis of Experiments*, 9e. Hoboken, NJ: Wiley.

Roberts, H.V. (1987). Data analysis for managers. *The American Statistician* 41 (4): 270–278.

Ronchi, F., Salmaso, L., De Dominicis, M. et al. (2017). Optimal designs to develop and support an experimental strategy on innovation of thermoforming production process. *Statistica* 2: 77–109.

Yandel, B.S. (1997). *Practical Data Analysis for Designed Experiments*. New York: Routledge.

Index

*End-to-End Data Analytics for Product Development: A Practical Guide for Fast Consumer Goods
Companies, Chemical Industry and Processing Tools Manufacturers*, First Edition.
Rosa Arboretti, Mattia De Dominicis, Chris Jones, and Luigi Salmaso.
© 2020 John Wiley & Sons Ltd. Published 2020 by John Wiley & Sons Ltd.
Companion website: www.wiley.com/go/salmaso/data-analytics-for-pd